Generis
PUBLISHING

A New Method for the Preparation of Hydrazides of Substituted Gammahydroxybutanoic Acids

Tariel V. Ghochikyan
Melanya Samvelyan

Title: **A New Method for the Preparation of Hydrazides of Substituted Gammahydroxybutanoic Acids**

ISBN: 979-8-89966-140-2

Author: Tariel V. Ghochikyan, Melanya Samvelyan

Cover image: www.pixabay.com

Publisher: Generis Publishing
Online orders: www.generis-publishing.com
Contact email: info@generis-publishing.com

A NEW METHOD FOR THE PREPARATION OF HYDRAZIDES OF SUBSTITUTED GAMMA-HYDROXYBUTANOIC ACIDS

Tariel V. Ghochikyan, Melanya A. Samvelyan

Faculty of Chemistry, Yerevan State University, 1 Alex Manoogian Str.

Contents

Abstract

A novel method for the synthesis of 4-hydroxybutyric acid hydrazides has been developed. The target compounds are produced by the method of hydrazinolis: via the reaction of 2-substituted (H, alkyl, aralkyl, allyl, etc.)-4-substituted-4-butanolides with 85 % hydrazine hydrate. Biological testing on white outbred mice demonstrated that the introduction of a hydroxypropyl group into the hydrazide structure confers new pharmacological properties, specifically hypotensive activity, which has not previously been observed in carboxylic acid hydrazides. The most active compounds were further tested on rats and cats, revealing a lethal dose (LD_{50}) range of 550–762 mg/kg, indicating superior activity compared to the reference drug "Dibazol".

To explore additional pharmacological potential, the antioxidant properties of these hydrazides were evaluated using competitive reactions with para-nitroso-N,N-dimethylaniline (PNDMA) as the acceptor. H radicals were generated by photolysis of hydrogen peroxide (10^{-3} mol/L) under UV light at 313 nm, and the radical reactivity was assessed by monitoring the rate of PNDMA decolorization. All tested compounds exhibited antioxidant activity, with several demonstrating effectiveness comparable to that of vitamin C. he hydrazinolysis of substituted 4-hydroxybutylmalonic acids in ethanol medium has also been investigated. It was demonstrated that the reaction pathway of 2-ethoxycarbonyl-4,4-disubstituted-4-butanolides with hydrazine hydrate under mild conditions (45–50 °C) is highly dependent on the nature of the substituent at the 2-position of the butanolide ring. The influence of these substituents on both the reaction mechanism and product yield was systematically studied.

The results show that when R^3 = H, the reaction predominantly yields dihydrazides of substituted malonic acids. In contrast, with alkyl-substituted butanolides, the same conditions favor intramolecular cyclization of the intermediate, leading to the formation of 4,4-disubstituted pyrazolidine-3,5-diones.

Physicochemical properties, including antioxidant activity, of the synthesized compounds were examined. UV-Vis absorption spectroscopy revealed substituent-dependent shifts in both the absorption wavelength and molar absorption coefficients, enabling determination of the transition types. Additionally, dynamic light scattering (DLS) measurements provided data on the mean hydrodynamic diameters of the synthesized molecules.

The antioxidant activity of the malonic acid derivatives was evaluated via a competitive reaction involving hydroxyl radicals and the free radical target para-nitroso-N, N-dimethylaniline (PNDMA), highlighting the potential of these compounds as antioxidant agents.

Introduction

Derivatives of organic acids, particularly hydrazides, are widely used as starting materials in fine organic synthesis. Despite their broad synthetic potential, comprehensive data on the biological activity of carboxylic acid hydrazides remain limited. Recent studies have shown that some naproxen-based acetohydrazides exhibit antitumor activity against human prostate carcinoma cell lines [1], while certain substituted benzoic acid hydrazides demonstrate similar effects [2]. Additionally, several hydrazone hydrazides have been reported to possess both antioxidant and antimicrobial properties [3].

Beyond their intrinsic biological activity, carboxylic acid hydrazides are also valuable intermediates in the synthesis of heterocyclic compounds, particularly oxadiazoles and 1,2,4-triazoles. The importance of hydrazides in these transformations is well established, especially considering the medical significance of various 1,2,4-triazole derivatives. Widely used pharmaceuticals such as Voriconazole, Triazolam, Fluconazole, Itraconazole, Furacilin, Alprazolam, and Estazolam are based on this heterocyclic scaffold. These compounds, notable for their synthetic accessibility and absence from natural sources, display a wide spectrum of biological activities.

Research into novel methods for synthesizing 1,2,4-triazoles remains active [4–7], driven by the increasing biological significance of these compounds. Studies have confirmed that various 1,2,4-triazole derivatives exhibit antimicrobial [8], antitumor [9,10], antibacterial [11–15], and fungicidal activities [16,17]. It has also been demonstrated that, with appropriate substitution patterns, triazoles can bind to the human adenosine A3 receptor and display insecticidal activity against *Tetranychus cinnabarinus* [18].

Despite the growing interest in hydrazide-based compounds, a literature review reveals an absence of data on γ-hydroxy acid hydrazides. This is particularly intriguing given that γ-hydroxybutyric acid (GHB), the simplest member of this class, plays a crucial role in the central nervous system. Its sodium salt is widely used in anesthesiology and ophthalmology. The incorporation of a GHB residue into biologically active molecules could therefore lead to the emergence of new and valuable pharmacological properties.

The scarcity of data on γ-hydroxy acid hydrazides is likely due to the limited availability and instability of the corresponding raw materials. γ-Hydroxy acids and their esters are known to be unstable, and GHB itself exists in equilibrium with its cyclic lactone form even in aqueous solution. Nevertheless, the unique structural and biological characteristics of GHB and its derivatives make them promising targets for further investigation.

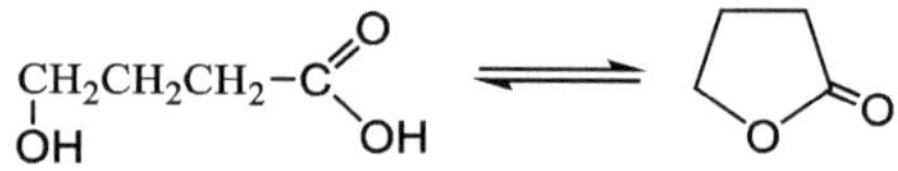

It is well established that dibasic organic acids and their derivatives have a wide range of industrial applications, including use as plasticizers, solvents, and intermediates in fine organic synthesis. Recent studies [20–23] have demonstrated that various derivatives of substituted glutaric acid exhibit diverse biological activities. Specifically, dihydrazides (compound 1) have shown significant hypotensive effects; mixed esters (compound 2) possess anti-inflammatory and psychotropic properties; allylglutaric acid imides (compound 3) act as anti-inflammatory agents; and the monosodium salt of glutaric acid amide (compound 4) displays antimineralocorticoid activity (Scheme 1).

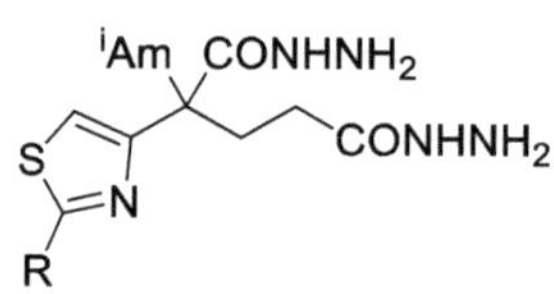

Compound 1

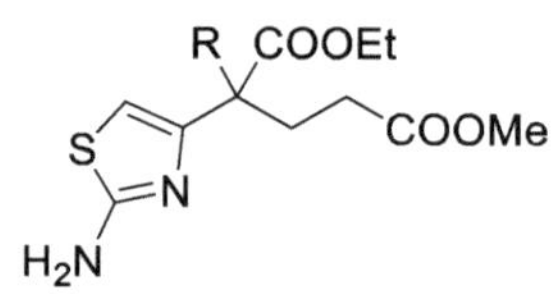

Compound 2

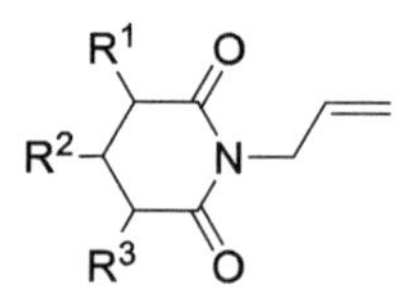

Compound 3

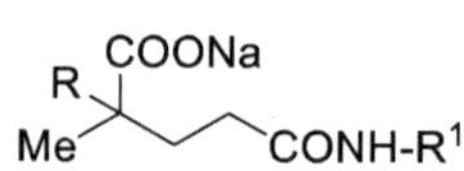

Compound 4

Scheme 1. Structure of some derivatives of dibasic acids

From the comparison of the various illustrated derivatives of dicarboxylic acids, it can be concluded that the type and intensity of biological activity are strongly dependent on the molecular structure of the compounds. The present study focuses on the synthesis, characterization, and investigation of malonic acid hydrazides bearing different substituents. A primary objective is to explore the relationship between structural modifications, physicochemical properties, and biological activity of the synthesized compounds.

The size of the synthesized compounds in solution was investigated using dynamic light scattering (DLS), a modern and increasingly relevant technique in pharmaceutical and nanomaterial research [24–26]. The hydrodynamic diameter of molecules in solution can serve as an important quality control parameter, especially for newly synthesized compounds exhibiting biological and antioxidant activity [27–29]. Although limited methods exist for measuring the size of small molecules in solution, DLS remains one of the most effective approaches, providing reliable measurements of particle size distributions [30,31]. In this study, DLS measurements were performed to determine the hydrodynamic diameters and size distributions of compounds 2a–e in aqueous solutions.

Another crucial property examined was the antioxidant activity of the synthesized compounds. Numerous exogenous compounds, including environmental pollutants and food additives, may enter the human body and negatively affect health and longevity [32–34]. The worsening of environmental conditions has been directly linked to an increase in oxidative stress—a pathological state caused by the excessive accumulation of free radicals. Oxidative stress plays a significant role in the onset and progression of cardiovascular, neurological, respiratory, and hematological disorders, as well as various ocular diseases. Due to their high reactivity, free radicals promote oxidative degradation of cellular macromolecules, ultimately leading to a range of pathological conditions. Antioxidants help mitigate these effects by neutralizing free radicals and protecting cellular components [35–39]. Consequently, the interaction between antioxidants and reactive species, particularly hydroxyl radicals, remains a subject of intense research in both biomedical and environmental fields [40–43].

One established method for synthesizing derivatives of dibasic acids involves the reaction of substituted 4-butanolides with nucleophiles such as hydrazine hydrate. According to literature reports, when the starting butanolide contains an ester group, ring opening occurs during the initial stage of the reaction, forming a hydrazide ester intermediate. Subsequent reaction with hydrazine leads to the formation of the corresponding dihydrazide [44].

In this study, we investigated the effect of various substituents on the mechanism and yield of this reaction. For this purpose, 2-ethoxycarbonyl-4,4-disubstituted-4-

butanolides were selected as starting materials, and hydrazine hydrate was used as the nucleophile (see Scheme 2 and Table 1).

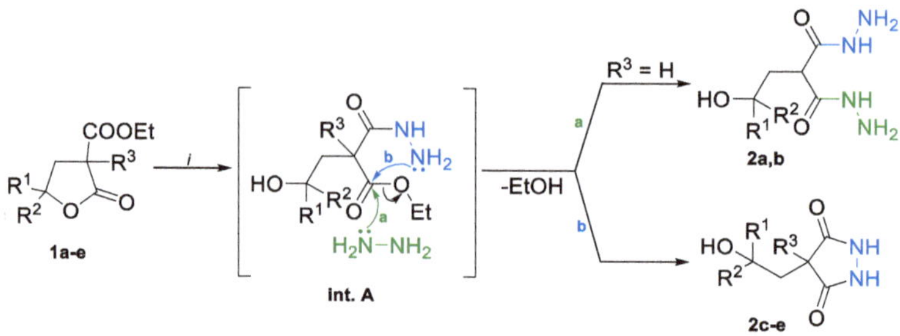

Scheme 2. Synthesis of derivatives of malonic acid. Reagents and conditions: (i) N_2H_4(aq), EtOH, 45-50 °C 2 h

Table 1. Synthesis of derivatives of malonic acid 2a-e

#	R^1	R^2	R^3	Yield %
2a	^{i}BuOCH$_2$	H	H	85
2b	Me	H	H	90
2c	^{i}BuOCH$_2$	H	Bu	79
2d	PrOCH$_2$	H	Bu	83
2e	Me	Me	^{i}Am	88

When R^3 = H, the reaction yields dihydrazides of substituted malonic acids. However, in the case of 2-alkyl-substituted butanolides, the same reaction conditions lead to an intramolecular cyclization of the intermediate, resulting in the formation of 4,4-disubstituted pyrazolidine-3,5-diones. To rationalize this behavior observed with 2-ethoxycarbonyl-2-alkyl-4-butanolides, we propose the following mechanism: hydrophobic interactions between the alkyl and hydroxypropyl groups located at the 2-position reduce the spatial distance between the ester and amide functionalities. This spatial proximity favors an intramolecular reaction between the hydrazinocarbonyl moiety and the ester group, ultimately leading to the formation of a more thermodynamically stable cyclic structure—namely, the pyrazolidine-3,5-dione.

All synthesized molecules were structurally characterized by ^{1}H NMR, ^{13}C NMR, UV-Vis spectroscopy, and LC-MS. The obtained data were in full agreement with the proposed structures.

This study focuses on the influence of alkyl substituents on the reaction pathway, the resulting physicochemical properties, and the antioxidant activity of the synthesized compounds. To the best of our knowledge, similar investigations on such systems have not been reported in the literature.

UV-vis spectra of compounds (2a–e) in aqueous solutions reveal the effect of structural changes on the spectral characteristics of these compounds (Fig. 1 and Table 2).

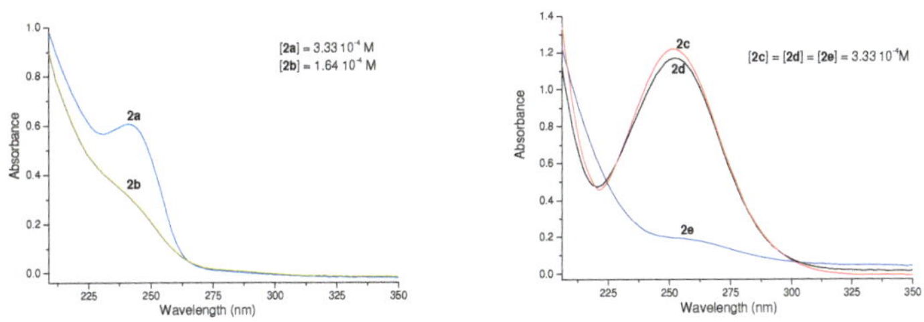

Fig. 1. UV-Vis spectra for compounds 2a-e

Table 2. UV-vis data of compounds (2a–e) in aqueous solutions

#	λ_{max}, nm	ε, $l \cdot mol^{-1} \cdot cm^{-1}$
2a	241	1819
2b	247(shoulder)	1514
2c	253	3532
2d	252	3661
2e	261(shoulder)	528

The absorption maxima and corresponding molar extinction coefficient values for the synthesized compounds in aqueous solutions are presented in Table 2. These spectral features are characteristic of $n \rightarrow \pi*$ transitions, primarily associated with the carbonyl functional groups present in the molecules. As observed from the data, both the hydrazide moiety and the pyrazolidine ring significantly influence the UV-Vis spectral characteristics. Specifically, absorption maxima at 252 and 261 nm are attributed to substituted pyrazolidine-3,5-diones, while lower wavelength absorptions at 241 and 247 nm correspond to compounds containing hydrazide groups.

The hydrodynamic sizes of the synthesized compounds were analyzed using dynamic light scattering (DLS). This technique measures time-dependent fluctuations in the intensity of light scattered by particles in solution at a fixed angle. The scattering angle is selected automatically by the DLS system software (Kalliope), which continuously monitors sample transmission. In all measurements performed, side scattering at 90° was selected by default.

The hydrodynamic diameter (d_H) of the particles is calculated using the Stokes–Einstein equation:

$$d_H = \frac{k_B T}{6\pi\eta D} \quad (1)$$

where k_B is the Boltzmann constant, η is the solvent viscosity, T is the absolute temperature, and D is the translational self-diffusion coefficient, obtained from the autocorrelation function of scattered light intensity fluctuations.

The data of mean hydrodynamic diameters, d_H, and polydispersity indexes (PDIs) of the sizes obtained for all solutions of compounds (2a-e) at 298 K are presented in Table 3.

Table 3. The hydrodynamic diameters, d_H, and polydispersity indexes (PDIs) of the sizes of solutions of compounds (2a–e) at 298 K

#	solvent	d_H, nm	PDIs , %
2a	EtOH	1.08±0.04	41.0
2b	EtOH	1.15±0.15	52.0
2c	EtOH	0.83±0.13	43.1
2d	H_2O	1.05±0.03	34.0
2e	H_2O	0.96±0.05	64.0

The intensity-weighted size distributions for some compounds are presented in supporting material.

The antioxidant activity of organic compounds (2a-e) was determined from the study of competitive reaction between hydroxyl radicals and free radical target p-nitroso-N, N-dimethyl aniline (PNDMA). The reaction with free radicals causes de-colorization of PNDMA and, by measuring the absorption at 440 nm it is possible to calculate the rate of reaction. The dependences of absorption of PNDMA from the period of radiation of hydrogen peroxide in the absence and presence of compounds (2a-e) are depicted in the supporting material. It is obvious that with the increase of concentrations of compounds (2a-e) the rate of competitive hydroxyl radical - PNDMA reaction decreases significantly, showing that these compounds have pronounced antioxidant properties.

It turned out, that increase of its concentration results in the decrease of the rate of decolorization of PNDMA, which means that compounds (2a-e) react with hydroxyl radicals and as a result, the competitive PNDMA-hydroxyl radical reaction becomes slower. From the slopes of the plots for different contents of compounds (2a-e) the rate of hydroxyl radical-PNDMA reaction is determined and the plots of slopes versus molar concentration of compounds (2a-e) are depicted in Fig. 2.

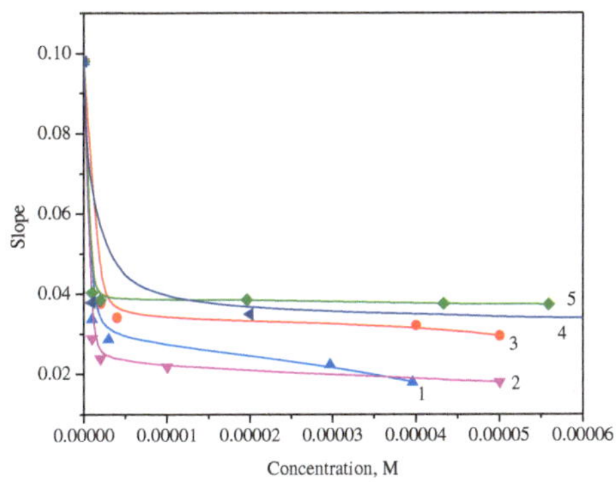

Fig. 2. Dependence of slopes of plots presented in supporting materials from the molar concentration. 1 – (2c); 2 – (2d); 3 – (2a); 4 – (2b); 5 – (2e)

The rate constants of compounds (2a-e) – hydroxyl radicals reaction are determined from the slopes of the plots of rate of the reaction versus concentration using (2) equation.

$$k_{OH+antioxidant} = 1.25 \cdot 10^{10} \cdot \frac{[PNDMA]}{[Antioxidant]} \cdot \left(\frac{W_1}{W_2} - 1\right) \quad (2)$$

The results are presented in Table 4.

Table 4. Rate constants of the reaction between (2a-e) and hydroxyl radicals

#	$k_{OH+antioxidant}$, L mol^{-1} sec^{-1}
2a	1.40×10^{8}
2b	0.73×10^{8}
2c	0.587×10^{8}
2d	0.344×10^{8}
2e	1.53×10^{8}

The data presented in Table 4 indicate that the rate of reaction between compounds (2a–e) and hydroxyl radicals varies significantly depending on the molecular structure. Among the tested compounds, the highest reactivity was observed for compound (2e), the reactivity decreasing in the following order: (2e) > (2a) > (2b) > (2c) > (2d).

For comparison, it is noteworthy that the rate constant for the reaction between hydroxyl radicals and the well-known antioxidant ascorbic acid, determined using the same experimental method, is 9.45×10^9 $L \cdot mol^{-1} \cdot s^{-1}$ [32]. This benchmark allows for an assessment of the relative antioxidant efficiency of the synthesized compounds.

Results and Discussion

Dihydrofuran-2(3H)-one (also known as butyrolactone or butan-4-olide) is the cyclic ester form of γ-hydroxybutyric acid (GHB). Similar cyclization reactions can occur with other γ-hydroxycarboxylic acids, although often under more rigorous conditions, yielding dihydrofuran-2(3H)-ones with varying substitution patterns. In our previous work, we synthesized a series of 2,4,4-trisubstituted-4-butanolides (compounds 1a–k) [51,52].

The corresponding hydrazides (compounds 2a–k) were subsequently synthesized [53] and subjected to preliminary biological screening. Several of these compounds displayed both **hypotensive** and **antioxidant** activity. Among them, compounds **2b** and **2d** were selected for detailed pharmacological studies in rats and cats.

In experiments conducted on anesthetized animals (Nembutal 50 mg/kg), the systemic arterial pressure (SAP) response to intravenous administration of the test compounds was assessed. In a separate series of experiments on anesthetized cats, the influence of the compounds on central hemodynamic parameters was investigated. SAP was continuously recorded in the common carotid artery. The **minute blood volume (MBV)** was evaluated using the **thermodelamination** method. Heart rate (HR) was derived from the electrocardiogram (lead II), and other cardiovascular indices including **total peripheral resistance (TPR)**, **stroke volume (SV)**, and **left ventricular work (LVW)** were calculated accordingly.

Acute toxicity studies were performed on white mice via intraperitoneal injection, following the Miller and Tainter method. Compound **2b** exhibited an LD_{50} of 762 mg/kg, while compound **2g** showed a higher toxicity with an LD_{50} of 550 mg/kg.

The test compounds were administered intravenously in aqueous solution at doses of **10, 20, 50, and 100 mg/kg**. **Compound 2b** produced a moderate and transient reduction in systolic blood pressure (SBP) at doses of 20 and 50 mg/kg, and a more pronounced hypotensive effect at 100 mg/kg (Table 2). In contrast, **compound 2g** demonstrated significantly higher hypotensive activity. Even at the lowest tested dose (10 mg/kg), it induced a marked reduction in SBP, and the hypotensive effect increased significantly with escalating doses (20 and 50 mg/kg).

These results suggest a strong structure-activity relationship, where specific substituents significantly influence both the **cardiovascular effects** and **toxicity profiles** of the synthesized hydrazides. Further structure–property investigations are ongoing to optimize therapeutic activity while minimizing toxicity.

Table 5. The effect of compounds **5b** and **5d** on the level of SBP (experiments on rats)

Comp. #	Dose *mg/kg*	Number experiences	Changes in SBP* after administration of the substance (in% to the initial) after a time, *min.*				
			5	15	30	45	60
5b	20	5	-21±6**	-13±5	-8±5	-3±4	-1±4
	50	5	-18±5**	-9±3	-8±4	-7±2	-5±2
	100	5	-26±6**	-16±7	-14±5	-13±5	-11±5
5d	20	5	-17±4**	-13±6	-9±3	-3±3	-3±3
	50	5	-25±7**	-14±4**	-12±4	-10±4	-6±2
	100	5	-35±8**	-27±8**	-19±6**	-18±6**	-11±5

* – "-" decrease, "+" increase SBP, ** – changes SBP statistically reliable

In experiments on anesthetized rats, compound 5d caused a lesser hypotensive effect than in experiments on rats. This can probably be explained by the unequal species-specific differences in sensitivity to this substance. The decrease in SBP after the introduction of 5d in experiments on cats is due to a decrease in OPSS, while the IOC increased slightly (Tables 3, 4). HR and LVEF after the introduction of 5d at doses of 25 and 50 mg/kg remained virtually unchanged. Thus, compounds 5b and 5d have pronounced hypotensive activity, while being low in toxicity (Tables 5, 6).

Table 6. The effect of compound **5d** on the level of SBP (experiments on white pedigree rats)

Dose *mg/kg*	Time from start experience *min*	Changes in SBP* after administration of the substance (in% to the initial) after a time, *min*					
		SBP	HR	MVB	TPR	SV	LVF
25	Initial data	170±20.2	148±12.2	686±42.6	15156±139.8	12.3±2.3	14103±96.2
	5	-1.2±0.4	-1.9±3.9	-38.7±6.4	+63.3±6.4	+38.4±4.3	-39.0±9.2
	15	-1.2±0.4	-1.9±4.2	-29.9±6.8	+40.65±5.8	-29.5±4.8	-39.0±9.2
	30	-1.8±0.6	-3.6±3.8	-17.6±8.9	+31.9±12.4	-16.2±3.6	-17.3±6.4
	45	-1.9±0.6	-4.6±3.6	-15.8±7.4	+19.6±7.9	-10.1±4.2	-12.2±6.9
	60	-2.0±0.5	-4.6±3.4	+24.8±12.6	-16.4±9.6	+12.8±6.4	+11.8±6.2

* – "-" decrease, "+" increase SBP

Table 7. The effect of compound **5d** on the level of SBP (experiments on white pedigree rats)

Dose *mg/kg*	Time from start experience *min*	Changes in SBP* after administration of the substance (in% to the initial) after a time, *min*					
		SBP	HR	MVB	TPR	SV	LVF
50	Initial data	140±20.6	140±25.3	292±42.6	31169±83.4	2.3±2.3	3577±64.3
	5	-6.4±3.6	-0.8±1.2	+4.1±3.8	-9.7±4.3	+15.3±6.2	+3.7±3.5
	15	-14.6±4.9	-0.8±1.2	+0.5±3.2	-16.8±3.7	+6.9±2.4	-18.3±6.8
	30	-14.6±4.9	-0.4±0.2	+6.3±2.4	-21.4±6.2	+3.4±1.8	-1.5±0.9
	45	-17.5±5.3	0	+16.2±4.6	-29.5±4.3	+13.9±3.3	-0.9±0.2
	60	-17.6±5.4	0	+12.4±6.2	-38.0±5.5	+12.5±3.8	+6.9±4.3

* – "-" decrease, "+" increase SBP

Table 8. Comparative effect of compound **5b** (experiments on pedigree pubescent rats)

(in comparable doses of 1/10 of LD_{50})

Dose *mg/kg*	Time from start experience *min*	Changes in SBP* after administration of the substance (in% to the initial) after a time, *min*					
		SBP	HR	MVB	TPR	SV	LVF
60	Initial data	125±15.3	200±13.3	184±58	50497±320	1.2±0.3	2931±305
	5	-4.4±1.3	-1.2±4.3	-6.7±7.3	+30.5±12.3	-7.2±5.8	-9.5±1.5
	15	-4.4±1.3	-5.5±6.1	-8.5±3.5	+40.2±8.7	-12.6±3.4	-10.5±4.2
	30	-8.7±2.9	-9.9±9.0	-22.0±9.8	+55.8±14.6	-18.7±9.8	-18.7±9.8
	45	-8.7±2.9	-9.9±9.0	-22.0±9.8	+55.8±14.6	-18.7±9.8	-18.7±9.8
	60	-12.0±4.0	-9.9±9.0	-30.6±12.4	+50.3±6.7	-18.7±6.4	-20.5±7.5

* – "-" decrease, "+" increase SBP

Table 9. Comparative effect of *Dibazole* (experiments on edigree pubescent rats)

(in comparable doses of 1/10 of LD_{50})

Dose *mg/kg*	Time from start experience *min*	Changes in SBP* after administration of the substance (in% to the initial) after a time, *min*					
		SBP	HR	MVB	TPR	SV	LVF
60	Initial data	140±4.6	200±18.5	328±12.6	32012±159.8	2.4±0.7	0.7±0.1
	5	-21.4±6.8	-10.6±8.3	-15.2±3.8	-27.3±6.2	-21.9±3.6	-30.9±5.6
	15	-7.8±2.4	-7.1±8.1	-23.8±1.6	-13.8±3.2	-29.8±5.4	-10.4±8.7
	30	-4.2±3.8	-1.3±0.9	-34.3±6.5	-1.9±1.8	-59.6±8.7	-12.1±8.7
	45	-1.5±2.2	+0.7±1.2	-19.4±5.4	36.7±6.4	-27.6±10.	-4.3±3.2
	60	+5.1±4.3	+2.4±1.3	-17.4±3.6	-63.5±9.8	+21.3±9.7	+9.9±7.8

* – "-" decrease, "+" increase SBP

To explore additional useful properties within the series of γ-hydroxy acid hydrazides, their antioxidant activity was evaluated using a competitive reaction method. In this assay, 4-nitroso-N, N-dimethylaniline (PNDMA) served as a competitive acceptor for hydroxyl radicals (·OH). The reactivity of hydroxyl radicals toward the synthesized compounds 5a–k was assessed by monitoring the rate of decolorization of PNDMA.

Hydroxyl radicals (·OH) were generated via photolysis of hydrogen peroxide (10^{-3} mol/L) under UV irradiation at λ = 313 nm. The rate of hydroxyl radical generation and its interaction with the test compounds was determined by measuring the change in optical density of PNDMA in distilled water in the presence of compounds 5a–k.

Kinetic data illustrating the influence of varying concentrations of the compounds on the optical density of PNDMA over time are presented in **Figures 3–5**. These results provide insight into the antioxidant capacity of the synthesized hydrazides and the structure–activity relationships influencing their ability to scavenge reactive oxygen species.

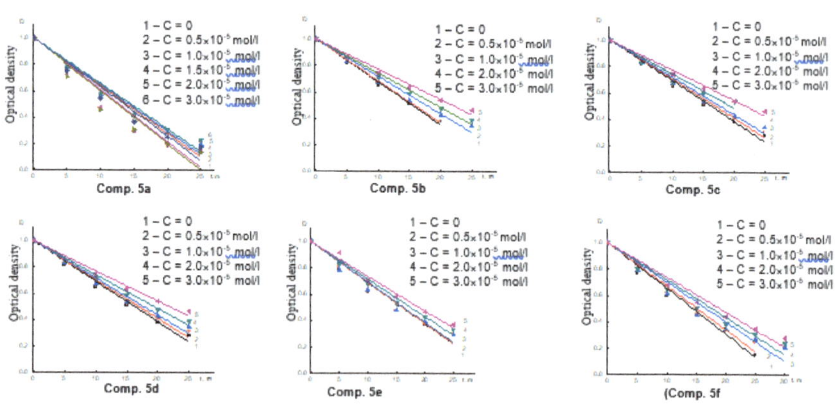

Figure 3. Dependence of optical density of PNDMA vs the time of UV irradiation (λ = 313 nm) at various concentrations of compounds 5a-f

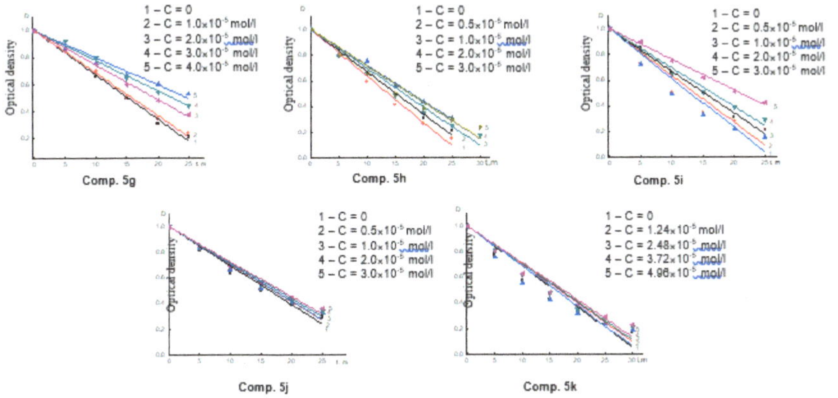

Figure 4. Dependence of optical density of PNDMA vs the time of UV irradiation (λ = 313 *nm*) at various concentrations of compounds **5g-k**.

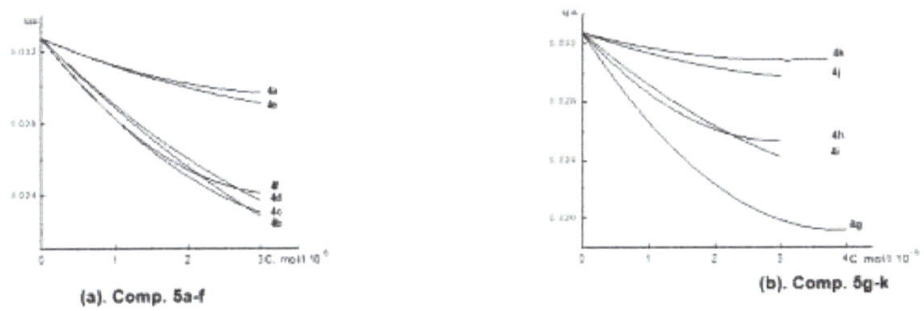

Figure 5. The dependence of the rate of PNDMA decolouration (in relative units) on concentration

Based on the experimental data (Fig. 3-5) according to [54-57], the rate constants were calculated using equation (1):

$$k_{OH+P} = 1.25\times10^{10}\times\frac{[PNDMA]}{[P]}\times\left[\frac{W_1}{W_2}-1\right]\left(\frac{L}{mol\times sec}\right) \qquad (1)$$

where 1.25x10^{10} is the rate constant of the interaction of $\dot{O}H$ radicals with PNDMA, [P] is the concentration of compounds 2a-k, W1 and W2 are the rates of PNDMA decolorization in distilled water and in the presence of the studied compounds, respectively. The following data were obtained for the rate constant of the reaction of the interaction of $\dot{O}H$ radicals and the studied substance (Table 7).

Table 10. The reaction rate constant of the interaction of $\dot{O}H$ radicals and the test compounds were obtained

Compound #	Reaction rate constant for interaction of $\dot{O}H$ radicals + test compound	Compound #	Reaction rate constant for interaction of $\dot{O}H$ radicals + test compound
5a	3.92×10^8	**5g**	6.2×10^8
5b	4.82×10^8	**5h**	2.7×10^8
5c	4.86×10^8	**5i**	4.0×10^8
5d	4.95×10^8	**5j**	1.71×10^8
5e	1.75×10^8	**5k**	0.965×10^8
5f	5.25×10^8	Ascorbic acid (control)	9.45×10^9

For the first time, a method for obtaining 4-hydroxybutanoic acid hydrazides has been proposed. It has been established that the target products are obtained on the basis of 2-substituted (H, alkyl, aralkyl, allyl, etc.) - 4,4-disubstituted-4-butanolides, which undergo hydrazinolysis with an 85% solution of hydrazine hydrate.

Preliminary screening has established that the newly synthesized compounds exhibit hypotensive activity. The selected, most active compounds were tested on rats and cats. It has been reliably established that the proposed compounds are superior in activity to the drug Dibazol used in medical practice.

The method of competitive reactions has established that the proposed compounds also exhibit significant antioxidant activity.

The most promising compounds were studied on rats. In particular, the effect of compounds 5b and 5g on the level of systemic arterial pressure (SAP) was studied in the usual way in experiments on anesthetized (Nembutal 50 mg/kg) animals. In a separate series of experiments on anesthetized cats, the effect of substances on central hemodynamics was studied; SAP was recorded in the common carotid artery. The minute blood volume (MBV) was studied by the thermodilation method. Heart rate (HR) was calculated from the electrocardiogram recorded in the 2nd standard lead. Total peripheral resistance (TPR), stroke volume (SV) and left ventricular work (LVW) were determined by calculation.

Acute toxicity was studied in experiments on white mice with intraperitoneal administration using the Miller and Teitner method. Acute toxicity of 5b was 762 mg/kg, and 5d – 550 mg/kg. The studied substances were administered intravenously in aqueous solution at doses of 10, 20, 50 and 100 mg/kg.

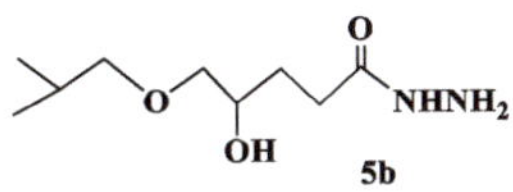

5b

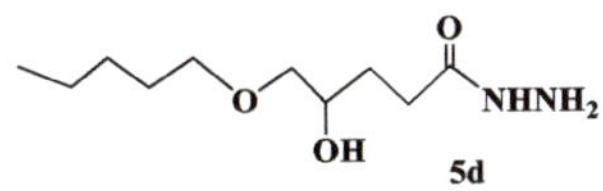

5d

In order to expand the arsenal of initial compounds, we have developed and improved the conditions for the synthesis of 2,4,4-trisubstituted-4-butanolides. Based on the latter, target hydrazides have been obtained, which are excellent initial compounds for the synthesis of various heterocyclic systems.

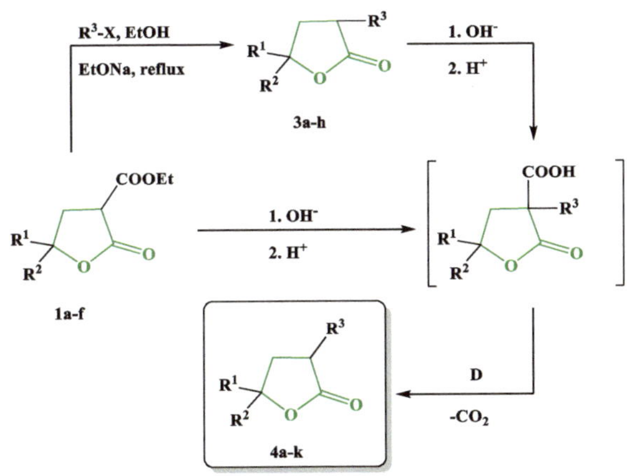

Scheme 3. Synthesis of 2,4,4-trisubstituted-4-butanolides

Table 8. 2,4,4–Trisubstituted-2-ethoxycarbonyl-4-butanolides

Comp.	R^1	R^2	R^3	Yield, %
3a	Izo-$C_4H_9OCH_2$	H	CH_2-C_6H_5	71
3b	n-$C_5H_{11}OCH_2$	H	CH_2-C_6H_5	74
3c	C_5H_{11} OCH_2	H	CH_2-CH=CH_2	85
3d	n-$C_3H_7OCH_2$	H	C_2H_5	82
3e	n-$C_3H_7OCH_2$	H	n-C_4H_9	79
3f	CH_3	H	izo-C_5H_{11}	84
3g	CH_3	CH_3	n-C_4H_9	86
3h	CH_3	H	izo-C_5H_{11}	84

Table 9. 3,5,5- trisubstituteddihydrofuran-2(3H)-one

Comp.	R^1	R^2	R^3	Yield, %
4a	H	H	H	87
4b	n-$C_3H_7OCH_2$	H	C_2H_5	93
4c	n-$C_3H_7OCH_2$	H	n-C_4H_9	91
4d	n-$C_4H_9OCH_2$	H	H	86
4e	izo-$C_4H_9OCH_2$	H	CH_2-C_6H_5	87
4f	izo-$C_4H_9OCH_2$	H	H	90
4g	izo-$C_4H_9OCH_2$	H	n-C_4H_9	85
4h	izo-$C_5H_{11}OCH_2$	H	CH_2-C_6H_5	91
4i	n-$C_5H_{11}OCH_2$	H	H	89
4j	n-C_5H_{11}	H	H	85
4k	n-C_5H_{11}	H	CH_2-CH=CH2	94
4l	CH_3	CH_3	n-C_4H_9	86
4m	CH_3	H	izo-C_5H_{11}	80

Table 9. 2,4,4-trisubstituted-γ-butanoic acid hydrazides

Comp.	R^1	R^2	R^3	Yield, %
5a	H	H	H	81
5b	n-$C_3H_7OCH_2$	H	C_2H_5	93
5c	n-$C_3H_7OCH_2$	H	H	93
5d	n-$C_3H_7OCH_2$	H	n-C_4H_9	83
5e	*izo*-$C_4H_9OCH_2$	H	H	90
5f	*izo*-$C_4H_9OCH_2$	H	n-C_4H_9	88
5g	*izo*-$C_4H_9OCH_2$	H	CH_2-C_6H_5	93
5h	*izo*-$C_5H_{11}OCH_2$	H	H	89
5i	*izo*-$C_5H_{11}OCH_2$	H	CH_2-C_6H_5	91
5j	$C_5H_{11}OCH_2$	H	H	89
5k	$C_5H_{11}OCH_2$	H	CH_2-CH=CH_2	89
5l	C_5H_{11}	H	H	85
5m	C_5H_{11}	H	CH_2-CH=CH_2	89
5n	C_5H_{11}	H	CH_2-C_6H_5	86
5o	CH_3	CH_3	n-C_4H_9	94
5p	CH_3	H	izo-C_5H_{11}	80
5q	CH_3	H	H	85
5r	CH_3	CH_3	izo-C_4H_9	71
5s	CH_3	CH_3	izo-C_5H_{11}	88
5t	CH_3	CH_3	n-C_4H_9	86
5u	CH_3	CH_3	C_2H_5	89

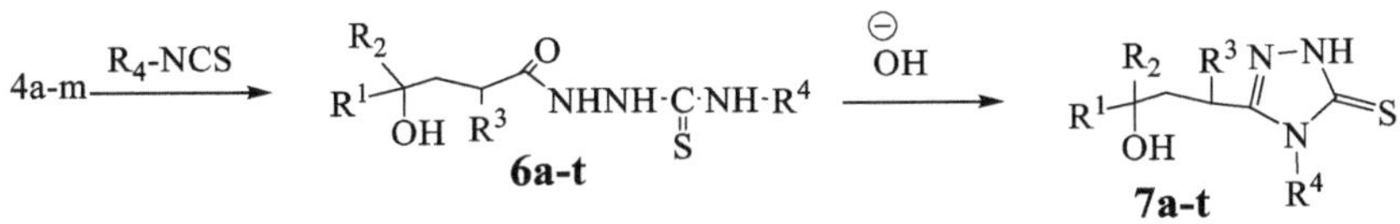

Scheme 4. Synthesis of trisubstituted-1,2,4-triazoles.

Table 10. 1,4-disubstituted thiosemicarbazides

Comp.	R^1	R^2	R^3	R^4	Yield, %
6a	n-$C_3H_7OCH_2$	H	H	C_6H_5	
6b	n-$C_3H_7OCH_2$	H	H	CH_2-CH=CH_2	
6c	n-$C_3H_7OCH_2$	H	n-C_4H_9	C_6H_5	
6d	n-$C_3H_7OCH_2$	H	n-C_4H_9	CH_2-CH=CH_2	
6e	n-$C_4H_9OCH_2$	H	H	C_6H_5	
6f	n-$C_4H_9OCH_2$	H	H	CH_2-CH=CH_2	
6g	izo-$C_5H_{11}OCH_2$	H	H	C_6H_5	
6h	izo-$C_5H_{11}OCH_2$	H	H	CH_2-CH=CH_2	
6i	izo-$C_5H_{11}OCH_2$	H	CH_2-C_6H_5	C_6H_5	
6j	C_5H_{11}	H	CH_2-C_6H_5	C_6H_5	
6k	CH_3	CH_3	n-C_4H_9	C_6H_5	76
6l	CH_3	CH_3	izo-C_5H_{11}	C_6H_5	75
6m	CH_3	CH_3	izo-C_5H_{11}	CH_2-CH=CH_2	81
6n	CH_3	CH_3	C_2H_5	C_6H_5	90
6o	CH_3	CH_3	C_2H_5	CH_2-CH=CH_2	82
6p	CH_3	H	izo-C_4H_9	CH_2-CH=CH_2	85
6q	CH_3	H	n-C_4H_9	CH_2-CH=CH_2	95
6r	CH_3	H	izo-C_5H_{11}	CH_2-CH=CH_2	94
6s	CH_3	H	n-C_4H_9	C_6H_5	78
6t	CH_3	H	izo-C_5H_{11}	C_6H_5	93

Table 11. Trisubstituted-1,2,4-triazoles

Comp.	R^1	R^2	R^3	R^4	Yield, %
7a	n-$C_3H_7OCH_2$	H	H	C_6H_5	
7b	n-$C_3H_7OCH_2$	H	H	CH_2-CH=CH_2	
7c	n-$C_3H_7OCH_2$	H	n-C_4H_9	C_6H_5	
7d	n-$C_3H_7OCH_2$	H	n-C_4H_9	CH_2-CH=CH_2	
7e	n-$C_4H_9OCH_2$	H	H	C_6H_5	
7f	n-$C_4H_9OCH_2$	H	H	CH_2-CH=CH_2	
7g	izo-$C_5H_{11}OCH_2$	H	H	C_6H_5	
7h	izo-$C_5H_{11}OCH_2$	H	H	CH_2-CH=CH_2	

7i	izo-$C_5H_{11}OCH_2$	H	CH_2-C_6H_5	C_6H_5	
7j	C_5H_{11}	H	CH_2-C_6H_5	C_6H_5	
7k	CH_3	CH_3	n-C_4H_9	C_6H_5	72
7l	CH_3	CH_3	izo-C_5H_{11}	C_6H_5	75
7m	CH_3	CH_3	izo-C_5H_{11}	CH_2-CH=CH_2	70
7n	CH_3	CH_3	C_2H_5	C_6H_5	74
7o	CH_3	CH_3	C_2H_5	CH_2-CH=CH_2	77
7p	CH_3	H	izo-C_4H_9	CH_2-CH=CH_2	69
7q	CH_3	H	n-C_4H_9	CH_2-CH=CH_2	73
7r	CH_3	H	izo-C_5H_{11}	CH_2-CH=CH_2	75
7s	CH_3	H	n-C_4H_9	C_6H_5	68
7t	CH_3	H	izo-C_5H_{11}	C_6H_5	80

Experimental part

Biological section

Compounds 2a-k [53] were subjected to preliminary biological screening and it was found that some of them, hydrazides, exhibit hypotensive and antioxidant activity. The most promising compounds were studied in rats and cats. In particular, the effect of compounds 2b and 2d on the level of systemic arterial pressure (SAP) was studied in the usual way in experiments on anesthetized (Nembutal 50 mg/kg) animals. In a separate series of experiments on anesthetized cats, the effect of the substances on central hemodynamics was studied; SAP was recorded in the common carotid artery. The minute blood volume (MBV) was studied by the thermodelamination method. The heart rate (HR) was calculated from the electrocardiogram recorded in the 2nd standard lead. Total peripheral resistance (TPR), stroke volume (SV) and left ventricular work (LV) were determined by calculation. Acute toxicity was studied in experiments on white mice with intraperitoneal administration according to the Miller and Teitner method. Acute toxicity of 4b was 762 mg/kg, and 2g – 550 mg/kg.

The studied substances were administered intravenously in an aqueous solution at doses of 10, 20, 50 and 100 mg/kg.

Compound 5b at doses of 20 and 50 mg/kg moderately and briefly reduced SBP, at a dose of 100 mg/kg, more pronounced hypotension was observed (Table 2). Compound 2g significantly exceeded compound 2b in hypotensive activity. Thus, even a dose of 10 mg/kg caused a distinct decrease in SBP. Increasing the dose to 20 and 50 mg/kg led to a significant increase in the hypotensive effect (Table 2).

Chemical section

All reagents were of analytical grade and were used as such or distilled prior the use. Starting 4-butanolides (1a-e) were prepared as previously reported [45,46].

^{1}H and ^{13}C NMR spectra were recorded on Varian Mercury-300 MHz in DMSO-CCl_4 mixture (1:3). Chemical shifts (δ) in ppm are reported as quoted relative to the residual signals of DMSO-d_6 (2.5 for ^{1}H NMR and 39.5 for ^{13}C NMR) as internal references. The coupling constants (J) are given in Hertz. TLC analysis was performed on silica gel "Silufol UV-254" plates. Melting points were determined on "Boetius" micro-heating stage. All reagents were of reagent grade and were used as such or distilled prior to use. Starting dihydrofuran-2(3*H*)-ones **2a-3k** was prepared as previously reported [20-21]. Melting points were determined on "Boetius" micro-heating stage

ESI-MS spectra were measured with a LC-MS 2020 Shimadzu.

The UV-vis absorption spectra were recorded with the Specord 50PC spectrophotometer using Aspect Plus Software. The path length used in absorption experiments was 1 cm. All measurements were performed at a constant temperature, 25.0 ± 0.3 Celsius, using the thermostatic device (LAUDA A100) attached to the spectrophotometer.

The dynamic light scattering (DLS) measurements were performed using the Litesizer 500 (Anton Parr, Austria) equipped with the Kalliope software. The "General" analysis model is used for determination of the particle size distribution chart and peak analysis. For determination of hydrodynamic diameter and polydispersity index the "Advanced" cumulant model with CONTIN algorithm is used. The measurement series were performed in quartz cuvettes at 298.15 K with 20 repetitions to verify reproducibility.

The optimal measured angle, number of runs, focus position, and optical filter settings were used as recommended by the equipment. Every sample was equilibrated before the measurement to avoid temperature gradients. The DLS experiments were reported with an uncertainty of less than 5%. An important factor for making good DLS measurements is using the right concentration. The concentration of particles must be high enough to produce a good scattering intensity. On the other hand, if the concentration is too high, then multiple scattering may occur, which can produce an erroneous result.

Therefore, to find out optimal concentration in DLS measurements, the measurements of several concentrations for each solution have been done. The solutions of (2a-e) compounds were prepared at least in three concentrations (0.01, 0.05, and 0.1 mol/L) in the deionized water and ethanol. It should be noted that the solubility of (2a-c) in water was less than those for (2d,e) compounds. Therefore for compounds (2a-c) ethanol was used as a solvent to obtain the same concentration range as it was for compounds (2d,e). The solvents were filtered through the 20 nm pore-size syringe filter (Whatman Anotop-25) before preparing the solutions. As an extra check, the particle-size measurements were carried out for the samples of the solvents before adding them to (2a-e) compounds. The absence of unwanted particles was in this way ensured.

Antioxidant activity of (2a-e) compounds was measured by PNDMA assay. In this method, the antioxidant properties are estimated by the study of kinetics of competitive reaction between hydroxyl radicals and PNDMA. Hydrogen peroxide is radiated by UV light at 313 nm causing the formation of hydroxyl radicals that react with PNDMA and as a result dye is decolorized. The rate of the reaction between hydroxyl radicals and PNDMA is determined from the absorption at 440 nm using photometric method. Addition of antioxidant affects this reaction due to the competitive reaction between (2a-e) compounds and hydroxyl radicals resulting in slower decolorization of radical target PNDMA. By the examination of rate of decolorization of dye the rate constant of reaction between (2a-e) compounds and hydroxyl radical may be determined according to the following equation [47-50].

$$k_{OH+antioxidant} = 1.25 \cdot 10^{10} \cdot \frac{[PNDMA]}{[Antioxidant]} \cdot \left(\frac{W_1}{W_2} - 1\right)$$

where [Antioxidant] and [PNDMA] are molar concentrations of antioxidant and PNDMA, respectively, $1.25 \cdot 10^{10}$ is the rate constant of reaction between hydroxyl radical and PNDMA ($l\,mol^{-1}sec^{-1}$), W_1 and W_2 are the slopes of plots of PNDMA absorption versus time of hydrogen peroxide radiation in the absence and presence of antioxidant, respectively.

General method for the preparation of compounds (2a-e)

0.05 mol of the corresponding 2,2,5-trisubstituted-4-butanolide solved in 100 ml EtOH is placed in Erlen-meyer flask. Then 7.4 g (0.125 mol) 85% of hydrazine hydrate was added. After solution was stirred and left for 2 h, then the reaction mixture was heated at 45-50 °C for 1 h. The ethanol was evaporated in vacuum and the crude solid was left for crystallization. The precipitated crystals washed with ether and dried.

2-(2-hydroxy-3-isobutoxypropyl)malonohydrazide (2a). Yield 11.3 g, 85%, m.p. 136-137 °C, R_f 0.49 (benzol: aceton: methanol - 1:5:15). ^{1}H NMR (300 MHz, DMSO/CCl_4 - 1/3), δ, p.p.m. (*J*, Hz): 8.77 br.s (1H, NH), 8.65 br.s (1H, NH), 4.64–3.54 m (5H, NH_2, OH), 3.53–3.40 m (1H, OCH), 3.31–3.20 m (3H, OCH_2, CH-C=O), 3.16 d (2H, OCH_2 $CHMe_2$, *J* 6.6), 2.01 ddd (1H, $HCCH_2CH$ J^1 13.3, J^2 9.9, J^3 3.0), 1.92–1.74 m (1H, $CHMe_2$), 1.56 ddd (1H, $HCCH_2CH$, J^1 14.1, J^2 9.9, J^3 4.6), 0.90 d

(6H, CH_3, J 6.7). ^{13}C NMR (75 MHz, DMSO/CCl_4 - 1/3), δ: 168.93, 168.60, 77.34, 75.13, 66.73, 46.05, 33.25, 27.80, 19.00.

LCMS [M+Na] 285.

2-(2-hydroxypropyl)malonohydrazide (2b). Yield 8.5 g, 90%, m.p. 131-132 °C, Rf 0.57 (aceton: hexane - 2:3). ^1H NMR (300 MHz, DMSO/CCl_4 -1/4), δ, p.p.m. (J, Hz): 8.80 br.s (1H, NH), 8.70 br.s (1H, NH), 4.13 s (4H, NH_2), 3.60–3.38 m (1H, OCH), 3.20 dd (1H, CH-C=O, J^1 9.4, J^2 5.2), 1.89 ddd ($1H^a$, CH_2, $J^1$13.3, J^2 9.4, J^3 3.7), 1.62 ddd ($1H^b$, CH_2, J^1 13.9, J^2 9.0, J^3 5.2), 1.07 d (3H, CH_3, J 6.1). ^{13}C NMR (75 MHz, DMSO/CCl_4 - 1/4) δ: 169.0, 168.8, 63.7, 46.6, 38.4, 23.5.

LCMS [M+1] 191, [M+Na] 213.

4-Butyl-4-(2-hydroxy-3-isobutoxypropyl)pyrazolidine-3,5-dione (2c). Yield 12 g, 83%, m.p. 146-147 °C, R_f 0.45 (benzol:aceton:methanol - 1:5:15). ^1H NMR (300 MHz, DMSO/CCl_4 - 1/3), δ, p.p.m. (J, Hz): 10.13 br.s (2H, NHC=O), 4.10 d (1H, OH, J 5.3), 3.74–3.60 m (1H, CHO), 3.23 dd ($1H^a$, CH_2O, J^1 13.4, J^2 9.5, J^3 5.6), 3.18–3.09 m (3H, CH_2O), 1.82 m (1H, $CHMe_2$), 1.72–1.63 m (2H, CH_2), 1.61–1.49 m (2H, CH_2), 1.30 m (2H, CH_2), 1.24–1.12 m (2H, CH_2), 0.95–0.86 m (9H,CH_3). ^{13}C NMR (75 MHz, DMSO/CCl_4 - 1/3) δ. 174.4, 173.4, 77.3, 75.0, 66.0, 49.6, 38.1, 35.2, 27.8, 25.4, 22.2, 19.0, 13.5.

LCMS [M+1] 287.

4-Butyl-4-(2-hydroxy-3-propoxypropyl)pyrazolidine-3,5-dione (2d). Yield 10.7 g, 79%, m.p. 126-128 °C, R_f 0.45 (benzol:aceton:methanol - 1:5:15). ^1H NMR (300 MHz, DMSO/CCl_4 - 1/3) δ, p.p.m. (J, Hz): 10.13 br.s (2H, NH), 4.11 d (1H, OH, J 5.4), 3.74–3.60 m (1H, CHO), 3.34 t (2H, OCH_2, J 6.6), 3.23 dd ($1H^a$, $CHCH_2O$, J^1 9.5, J^2 5.7), 3.15 dd ($1H^b$, $CHCH_2O$, J^1 9.5, J^2 5.7), 1.74–1.65 m (2H, CH_2), 1.63–1.48 m (4H, CH_2), 1.37–1.24 m (2H, CH_2), 1.24–1.10 m (2H, CH_2), 0.91 t (3H, CH_3, J 7.3), 0.90 t (3H, CH_3, J 7.3). ^{13}C NMR (75 MHz, DMSO), δ. 174.5, 173.4, 74.9, 72.1, 66.1, 49.7, 38.2, 35.3, 25.4, 22.4, 22.2, 13.5, 10.3.

LCMS [M+1] 273, [M+Na] 295.

4-(2-Hydroxy-2-methylpropyl)-4-isopentylpyrazolidine-3,5-dione (2e). Yield 10.6 g, 88%, m.p. 87-88 °C, R_f 0.60 (aceton: hexane - 2:3). ^1H NMR (300 MHz, DMSO/CCl_4 - 1/3) δ, p.p.m. (J, Hz): 8.47 br.s (2H, NH), 4.10 br.s (1H, OH), 2.87 d (1H, CH_2, J 13.3), 1.98–1.80 m (2H, CH_2), 1.74–1.60 m (1H, CH_2), 1.59–1.40 m (4H,

CH_2, CH_3), 1.32 s (3H, CH_3), 1.24–1.08 m (1H, CH_2), 1.08–0.94 m (1H, CH_2), 0.89 d (6H, CH_3, *J* 6.6). ^{13}C NMR (75 MHz, DMSO/CCl_4 - 1/3) δ, pp: 175.5, 167.5, 81.6, 55.5, 41.6, 35.0, 33.1, 29.1, 27.8, 27.5, 22.0, 22.0.

LCMS [M+1] 243, [M+Na] 265.

2,4,4–Trisubstituted-2-ethoxycarbonyl-4-butanolides (3a-h). 20 *ml* of absolute ethyl alcohol and 2.3 *g* (0.1 *mol*) of sodium metal are placed in a dry three-necked flask equipped with a stirrer, reflux condenser and dropping funnel. After dissolution and cooling, 0.1 *mol* of the corresponding 4,4-disubstuted-2-ethoxycarbonylbutanolide is added dropwise. The mixture is stirred for 15 *min* and 0.11 *mol* of the corresponding halogenide is added dropwise, stirred for 2 *h* without heating and at 75-80 °C until neutral reaction of the medium. After distillation of the ethyl alcohol the residue is cooled and acidified (HCl) water is added to pH 2-3; extracted with ether, the extracts are washed with water and dried with anhydrous magnesium sulfate. After distilling off the solvent, the residue is distilled.

Ethyl 3-benzyl-5-(isobutoxymethyl)-2-oxotetrahydrofuran-3-carboxylate (3a). Yield 71%, b.p. 160-161 °C/ 1Torr, n_D^{20} 1.5120, d_4^{20} 1.1439. Found, %: C 68.30; H 7.75. $C_{19}H_{26}O_5$. Calculated, %: C 68.24; H 7.84.

Ethyl 3-benzyl-5-(pentyloxy)methyl)-2-oxotetrahydrofuran-3-carboxylate (3b). Yield 74%, b.p. 167-168 °C/ 1Torr, n_D^{20} 1.5109, d_4^{20} 1.1175. Found, %: C 68.85; H 8.05. $C_{20}H_{28}O_5$. Calculated, %: C 68.94; H 8.10.

Ethyl 3-allyl-2-oxo-5-pentyltetrahydrofuran-3-carboxylate (3c). Yield 89%, b.p. 112-113 °C/ 1Torr, n_D^{20} 1.4675, d_4^{20} 1.0443. Found, %: C 68.85; H 8.05. $C_{15}H_{24}O_4$. Calculated, %: C 67.14; H 9.01.

Ethyl 3-ethyl-2-oxo-5-(propoxymethyl)tetrahydrofuran-3-carboxylate(3d). Yield 83%, b.p. 109-111 °C/ 1Torr, n_D^{20} 1.4515, d_4^{20} 1.0245. Found, %: C 60.50; H 8.10. $C_{13}H_{22}O_5$. Calculated, %: C 660.45; H 9.18.

Ethyl 3-butyl-5-(propoxymethyl)-2-oxotetrahydrofuran-3-carboxylate(3e). Yield 79%, b.p. 115-116 °C/ 1Torr, n_D^{20} 1.4525, d_4^{20} 1.0193. Found, %: C 64.05; H 9.35. $C_{16}H_{28}O_5$. Calculated, %: C 63.97; H 9.45.

Ethyl 3-butyl-2-oxo-5-methyl-tetrahydrofuran-3-carboxylate(3f). Yield 85%, b.p.117-118 °C/ 1Torr, n_D^{20} 1.4520, d_4^{20} 1.0200. Found, %: C 63.00; H 9.10. $C_{15}H_{26}O_5$. Calculated, %: C 62.91; H 9.15.

Ethyl 3-butyl-5,5-dimethyl-2-oxotetrahydrofuran-3-carboxylate (3g). Yield 85%, b.p. 96-97 °C/ 1Torr, n_D^{20} 1.4505, d_4^{20} 1.0258. Found, %: C 64.50; H 9.10. $C_{13}H_{22}O_4$. Calculated, %: C 64.44; H 9.15.

Ethyl 3-isopentyl-5-methyl-2-oxotetrahydrofuran-3-carboxylate (3h). Yield 85%, b.p. 99 °C/ 1Torr, n_D^{20} 1.4460, d_4^{20} 1.0210. Found, %: C 64.50; H 9.10. $C_{13}H_{22}O_4$. Calculated, %: C 64.44; H 9.15.

2,4,4-Trisubstituted butanolides (4a-p). 0.05 *mol* of the corresponding 2,4,4-trisubstituted-2-ethoxycarbonylbutanolide is added dropwise to a solution of sodium hydroxide [7 *g* (0.175 *mol*) of sodium hydroxide in 16 *ml* of water) and 0.5 *ml* of catamine AB; the whole is stirred for 1 *h* at 20-25 °C and 2*h* at 55-60°C. After cooling, the mixture is acidified with conc. hydrochloric acid to pH 2-3, the product is extracted with ether, washed with water and dried with anhydrous magnesium sulfate. After distilling off the solvent, the residue is subjected to decarboxylation by heating at 150-200°C and pressure of 15-20 mmHg, the residue is distilled.

3-Ethyl-5-(propoxymethyl)dihydrofuran-2(3*H*)-one(4a). Yield 88%, b.p. 94-95°C/ 1 Torr, n_D^{20} 1.4403, d_4^{20} 0.9345. Found, %: C 64.55; H 9.74. $C_{10}H_{18}O_3$. Calculated, %: C 64.49; H 9.74.

5-(Isobutoxymethyl)dihydrofuran-2(3*H*)-one (4b). Yield 75%, b.p. 92-93 °C/ 2Torr, n_D^{20} 1.4445, d_4^{20} 1.0151. Found, %: C 62.85; H 9.25. $C_9H_{16}O_3$. Calculated, %: C 62.77; H 9.36.

3-Butyl-5-(izobutoxymethyl)dihydrofuran-2(3*H*)-one(4c). Yield 91%, b.p. 100-101 °C / 1Torr, n_D^{20} 1.4425, d_4^{20} 0.9301. Found, %: C 62.70; H 9.40. $C_9H_{16}O_3$. Calculated, %: C 62.77; H 9.36.

5-((Pentyloxy)methyl)dihydrofuran-2(3*H*)-one (4d). Yield 71%, b.p. 98 °C/ 2Torr, n_D^{20} 1.4475, d_4^{20} 1.0075. Found, %: C 64.55; H 9.70. $C_{10}H_{18}O_3$. Calculated, %: C 64.49; H 9.74.

5-Pentyldihydrofuran-2(3*H*)-one (3e). Yield 98%, b.p. 81-82 °C/ 1Torr, n_D^{20} 1.4460, d_4^{20} 0.9502. Found, %: C 69.25; H 10.25. $C_9H_{16}O_2$. Calculated, %: C 69.19; H 10.32.

3-Butyl-5-(propoxymethyl)dihydrofuran-2(3*H*)-one (4f). Yield 90%, b.p. 105-106 °C / 1 Torr, n_D^{20} 1.4430, d_4^{20} 1.1193. Found, %: C 67.35; H 10.40. $C_{12}H_{22}O_3$. Calculated, %: C 67.26; H 10.35.

3-Benzyl-5-(isobutoxymethyl)dihydrofuran-2(3*H*)-one (4g). Yield 73%, b.p. 141-142 °C/ 1Torr, n_D^{20} 1.5020, d_4^{20} 1.0479. Found, %: C 73.20; H 8.50. $C_{16}H_{22}O_3$. Calculated, %: C 73.25; H 8.45.

3-Benzyl-5-((isopentyloxy)methyl)dihydrofuran-2(3*H*)-one (4h). Yield 75%, b.p. 157-158 °C/ 1Torr, n_D^{20} 1.5009, d_4^{20} 1.0373. Found, %: C 73.95; H 8.70. $C_{17}H_{24}O_3$. Calculated, %: C 73.88; H 8.75.

3-Allyl-5-pentyldihydrofuran-2(3*H*)-one (4i). Yield 91%, b.p. 95-96 °C/ 1Torr, n_D^{20} 1.4580, d_4^{20} 0.9442. Found, %: C 73.50; H 10.20. $C_{12}H_{20}O_2$. Calculated, %: C 73.43; H 10.27.

3-Butyl-5,5-dimethyldihydrofuran-2(3*H*)-one (4j). Yield 80%, b.p. 71-72 °C/ 1Torr, n_D^{20} 1.4400, d_4^{20} 0.9283. Found, %: C 70.50; H 10.60. $C_{10}H_{18}O_2$. Calculated, %: C 70.55; H 10.66.

3-Isopentyl-5-methyldihydrofuran-2(3*H*)-one (4k). Yield 79%, b.p. 72-73 °C/ 2Torr, n_D^{20} 1.4410, d_4^{20} 0.9365. Found, %: C 70.60; H 10.70. $C_{10}H_{18}O_2$. Calculated, %: C 70.55; H 10.66.

General method for the preparation Hydrazides of 2,4-disubstituted-4-hydroxypentanoic acids (5a-w). A mixture of 0.05 *mol* of 2,4,4-trisubstituted-4-pentanolide, 3 *g* (0.06 *mol*) of 85% hydrazine hydrate in 10 *ml* of ethanol is heated in a boiling water bath for 2 *h* and ethanol is distilled off. The crystalline residue is washed with ether and dried.

2-Ethyl-4-hydroxy-5-propoxypentanehydrazide (5a). Yield 93%, mp. 119-120 °C. ^{1}H NMR (300 MHz, DMSO/CCl_4 - 1/3) δ 8.70 (s, 1H, NHC=O), 3.91 (br.s, 3H, OH, NH_2), 3.60 – 3.47 (m, 1H, $\underline{CH}$OH), 3.35 (t, *J* = 6.6 Hz, 2H, $O\underline{CH_2}CH_2CH_3$), 3.23 (dd, *J* = 5.4, 1.3 Hz, 2H, $\underline{CH_2}$CHOH), 2.13 (dtd, *J* = 8.8, 7.0, 5.3 Hz, 1H, CHC=O), 1.62 – 1.33 (m, 6H, CH_2), 0.91 (t, *J* = 7.4 Hz, 3H, CH_3), 0.82 (t, *J* = 7.4 Hz, 3H,CH_3). ^{13}C NMR (75 MHz, DMSO/CCl_4 - 1/3) δ 174.6, 74.9, 72.1, 67.3, 41.9, 36.3, 24.7, 22.4, 11.4, 10.2. Found, %: C 55.10; H 10.10; N 12.90. $C_{10}H_{22}N_2O_3$. Calculated, %: C 55.02; H 10.16; N 12.83.

4-Hydroxy-5-isobutoxypentanehydrazide (5b). Yield 90%, m.p. 65-66 °C. ^{1}H NMR (300 MHz, DMSO/CCl_4 - 1/3) δ 9.62 (s, 0.4H, NHC=O), 8.77 (s, 0.6H, NHC=O), 4.25 (dd, *J* = 11.9, 4.8 Hz, 1H, OH), 3.90 (s, 2H, NH_2), 3.71 – 3.42 (m, 1H, CHO), 3.33 – 3.06 (m, 4H, OCH_2), 2.38 – 2.03 (m, 2H, CH_2C=O), 1.97 – 1.64 (m, 2H, CH_2), 1.59 – 1.32 (m, 1H, CH), 0.89 (d, *J* = 6.7 Hz, 6H, $CH(\underline{CH_3})_2$). ^{13}C NMR (75 MHz,

DMSO/CCl_4 - 1/3) δ 172.1, 170.7, 77.4, 75.0, 68.5, 68.3, 29.7, 29.4, 29.3, 27.9, 19.1. Found, %: C 53.00; H 9.80; N 13.80. $C_9H_{20}N_2O_3$. Calculated, %: C 52.92; H 9.87; N 13.71.

2-(2-hydroxy-3-isobutoxypropyl)hexanehydrazide (5c). Yield 85%, mp 126-127 °C. ^{1}H NMR (300 MHz, DMSO/CCl_4 - 1/3) δ 8.67 (s, 1H, NHC=O), 3.86 (br.s, 3H, OH, NH_2), 3.64 – 3.43 (m, 1H, CHOH), 3.29 – 3.18 (m, 2H, OCH_2CHO), 3.16 (d, J = 6.6 Hz, 2H, OCH_2CHMe_2), 2.19 (dtd, J = 9.4, 6.9, 4.8 Hz, 1H, CHC=O), 1.91 – 1.75 (m, 1H, $CHMe_2$), 1.57 – 1.41 (m, 3H, CH_2), 1.40 – 1.25 (m, 3H, CH_2), 1.24 – 1.10 (m, 2H, CH_2), 0.95 – 0.84 (d, J = 6.6 Hz, 6H, $CH(CH_3)_2$, d, J = 7.3 Hz, 3H, CH_2CH_3). ^{13}C NMR (75 MHz, DMSO/CCl_4 - 1/3) δ 174.7, 77.3, 75.0, 67.3, 40.3, 36.6, 31.5, 28.9, 27.8, 22.1, 19.9, 13.6. Found, %: C 60.05; H 10.75; N 10.85. $C_{13}H_{28}N_2O_3$. Calculated, %: C 59.97; H 10.84; N 10.76.

4-Hydroxy-5-(pentyloxy)pentanehydrazide (5d). Yield 89%, m.p. 71-72 °C. ^{1}H NMR (300 MHz, DMSO/CCl_4 - 1/3) δ 9.62 (br.s, 0.15H, NHC=O), 8.77 (br.s, 0.8H, NHC=O), 4.26 (d, J = 4.7 Hz, 0.8H, OH), 4.22 (d, J = 4.9 Hz, 0.15H, OH), 3.89 (br.s, 2H, NH_2), 3.65 – 3.46 (m, 1H, CHO), 3.38 (t, J = 6.6 Hz, 2H, CH_2CH_2O), 3.26 (dd, J = 9.5, 5.8 Hz, $1H^a$, $CHCH_2O$), 3.19 (dd, J = 9.5, 5.6 Hz, $1H^b$, $CHCH_2O$), 2.30 – 2.05 (m, 2H, CH_2CH_2C=O), 1.73 (dddd, J = 13.8, 8.0, 7.3, 3.5 Hz, $1H^a$, CH_2CH_2C=O), 1.59 – 1.42 (m, 2H, CH_2CH_2O, $1H^b$, CH_2CH_2C=O), 1.39 – 1.24 (m, 4H, $CH_2CH_2CH_3$), 0.97 – 0.85 (m, 3H, CH_3). ^{13}C NMR (75 MHz, DMSO/CCl_4 - 1/3) δ 172.1, 170.8, 74.8, 70.5, 68.5, 68.3, 29.7, 29.4, 28.9, 27.8, 22.0, 13.7. Found, %: C 55.05; H 10.10; N 12.85. $C_{10}H_{22}N_2O_3$. Calculated, %: C 55.02; H 10.16; N 12.83.

4-Hydroxynonanehydrazide (5e). Yield 85%, m.p. 81-82 °C. ^{1}H NMR (300 MHz, DMSO/CCl_4 - 1/3) δ 9.60 (br.s, 0.25H, NHC=O), 8.74 (br.s, 0.75H, NHC=O), 4.18 – 3.98 (m, 1H, OH), 3.88 (br.s, 2H, NH_2), 3.53 – 3.20 (m, 1H, CHO), 2.32 – 2.05 (m, 2H, CH_2C=O), 1.75 – 1.57 (m, $1H^a$, CH_2), 1.57 – 1.12 (m, 9H, CH_2), 0.90 (t, J = 6.9 Hz, 3H, CH_3). ^{13}C NMR (75 MHz, DMSO/CCl_4 - 1/3) δ 172.2, 170.7, 69.3, 69.1, 37.1, 32.7, 32.7, 31.5, 30.0, 29.7, 24.9, 22.1, 13.7. Found, %: C57.45 ; H 10.65; N 14.95. $C_9H_{20}N_2O_2$. Calculated, %: C 57.42; H 10.71; N 14.88.

2-(2-hydroxy-3-propoxypropyl)hexanehydrazide (5f). Yield 83%, mp 122-123 °C. ^{1}H NMR (300 MHz, DMSO/CCl_4 - 1/3) δ 8.68 (s, 1H, NHC=O), 3.89 (br.s, 3H, OH, NH_2), 3.61 – 3.47 (m, 1H, CHOH), 3.35 (t, J = 6.6 Hz, 2H, OCH_2CH_2), 3.23 (m, 2H, OCH_2CHOH), 2.18 (dtd, J = 9.4, 7.0, 4.9 Hz, 1H, CHC=O), 1.65 – 1.44 (m, 5H, CH_2), 1.42 – 1.24 (m, 3H, CH_2), 1.24 – 1.12 (m, 2H, CH_2), 0.96 – 0.85 (m, 6H, CH_3). ^{13}C NMR (75 MHz, DMSO/CCl_4 - 1/3) δ 174.7, 87.2, 74.8, 72.0, 67.3, 40.3, 36.7, 31.5,

28.9, 22.3, 22.1, 13.6, 10.2. Found, %: C 58.51; H 10.64; N 11.37. $C_{12}H_{26}N_2O_3$. Calculatrd, %: C 58.61; H 10.60; N 11.49.

2-Benzyl-4-hydroxy-5-isobutoxypentanehydrazide (5g). Yield 87%, m.p. 112 °C. ^{1}H NMR (300 MHz, DMSO/CCl$_4$ - 1/3) δ 8.61 (d, *J* = 7.0 Hz, 1H, NH), 7.32 – 7.05 (m, 5H$_{arom}$), 4.15 (d, *J* = 4.5 Hz, 0.5H, OH), 3.98 (d, *J* = 5.2 Hz, 0.5H, OH), 3.85 (br..s, 2H, NH$_2$), 3.65 – 3.37 (m, 1H, CHO), 3.29 – 3.16 (m, 2H, CH$_2$O), 3.14 (dd, *J* = 6.4, 3.0 Hz, 2H, CH$_2$O), 2.86 (ddd, *J* = 19.3, 12.9, 8.3 Hz, 1H, CHC=O), 2.72 – 2.51 (m, 2H, CH$_2$Ph), 1.88 – 1.75 (m, 1H, <u>CH</u>(CH$_3$)$_2$), 1.71 (ddd, 0.5H^a, CH<u>CH$_2$</u>CH), 1.62 – 1.52 (m, 1H^b, CH<u>CH$_2$</u>CH), 1.28 (ddd, *J* = 13.6, 10.4, 3.2 Hz, 0.5H^a, CH<u>CH$_2$</u>CH), 0.88 (dd, *J* = 6.7, 1.4 Hz, 6H, 2CH$_3$). ^{13}C NMR (75 MHz, DMSO/CCl$_4$ - 1/3) δ 173.9, 173.4, 139.7, 139.6, 137.1, 128.5, 128.5, 127.5, 127.4, 125.2, 125.2, 77.3, 75.3, 74.9, 67.3, 66.6, 42.3, 41.9, 38.6, 37.7, 36.3, 36.2, 27.8, 19.0. Found, %: C 65.30; H 8.85; N 9.60. $C_{16}H_{26}N_2O_3$. Calculated, %: C 65.28; H 8.90; N 9.52.

2-Benzyl-4-hydroxy-5-(isopentyloxy)pentanehydrazide (5h). Yield 91%, m.p. 110 °C. ^{1}H NMR (300 MHz, DMSO/CCl$_4$ - 1/3) δ 8.61 (br.s, 1H, NHC=O), 7.32 – 6.99 (m, 5H$_{arom}$), 3.97 (br.d, *J* = 4.7 Hz, 1H, OH), 3.82 (br.s, 2H, NH$_2$), 3.66 – 3.50 (m, 1H, CHO), 3.39 (t, *J* = 6.7 Hz, 2H, CH$_2$<u>CH$_2$</u>O), 3.22 (dd, *J* = 5.4, 2.8 Hz, 2H, CH<u>CH$_2$</u>O), 3.30 – 3.09 (m, 1H, CHC=O), 2.82 (dd, *J* = 13.3, 9.0 Hz, 1H^a, CH<u>CH$_2$</u>CH), 2.64 (dd, *J* = 13.3, 5.6 Hz, 1Hb, CH<u>CH$_2$</u>CH), 1.77 – 1.64 (m, 1H, <u>CH</u>(CH$_3$)$_2$), 1.64 – 1.48 (m, 2H, <u>CH$_2$</u>Ph), 1.48 – 1.31 (m, 2H, <u>CH$_2$</u>CH(CH$_3$)$_2$), 0.90 (d, *J* = 6.6 Hz, 5H, CH(<u>CH$_3$</u>)$_2$), 0.87 (d, *J* = 6.7 Hz, 1H, CH(<u>CH$_3$</u>)$_2$). ^{13}C NMR (75 MHz, DMSO/CCl$_4$ - 1/3) δ 173.9, 139.7, 128.5, 127.4, 125.2, 75.7, 74.9, 74.8, 68.7, 67.3, 42.3, 38.0, 37.7, 36.3, 35.4, 34.3, 25.6, 24.4, 22.3. Found, %: C 66.25; H 9.10; N 9.18. $C_{17}H_{28}N_2O_3$. Calculated, %: C 66.20; H 9.15; N 9.08.

2-Allyl-4-hydroxynonanehydrazide (5i). Yield 94%, m.p. 136-137 °C. ^{1}H NMR (300 MHz, DMSO/CCl$_4$ - 1/3) δ 8.70 (s, 1H, NH), 5.69 (ddt, *J* = 16.8, 10.1, 6.8 Hz, 1H, CH=), 5.03 – 4.94 (m, 1H^a, =CH$_2$), 4.94 – 4.88 (m, 1H^b, =CH$_2$), 4.00 – 3.69 (m, 3H, NH2, OH), 3.37 (br.s, 1H, CHO), 2.37 – 1.98 (m, 3H, =CH<u>CH$_2$CH</u>), 1.54 (ddd, *J* = 15.0, 8.3, 6.8 Hz, 1H^a, CH$_2$), 1.46 – 1.12 (m, 9H, CH$_2$), 0.90 (t, *J* = 6.9 Hz, 3H, CH$_3$). ^{13}C NMR (75 MHz, DMSO/CCl$_4$ - 1/3) δ 174.2, 136.1, 115.3, 67.9, 40.3, 39.4, 37.2, 36.0, 31.4, 24.7, 22.1, 13.7. Found, %: C 63.20; H 10.55; N 12.35. $C_{12}H_{24}N_2O_2$. Calculated, %: C 63.12; H 10.59; N 12.27.

2-(2-Hydroxy-2-methylpropyl)hexanehydrazide (5j). Yield 86%, m.p. 105-106 °C. Found, %: C 59.30; H 11.00; N 13.95. $C_{10}H_{22}N_2O_2$. Calculated, %: C 59.37; H 10.96; N 13.85.

^1H NMR (300 MHz, DMSO/CCl_4 - 1/3) δ 8.67 (br.s, 1H, NHC=O), 3.84 (br.s, 2H, NH_2), 3.55 (br.s, 1H, OH), 2.28 – 2.15 (m, 1H, CHC=O), 1.84 (dd, J = 13.9, 9.4 Hz, $1H^a$, CH_2), 1.57 – 1.37 (m, $1H^a$, CH_2), 1.36 – 1.09 (m, 6H, CH_2), 1.05 (d, J = 6.3 Hz, 6H, $C(CH_3)_2$), 0.89 (t, J = 7.0 Hz, 3H, $CH_2\underline{CH_3}$).

^{13}C NMR (75 MHz, DMSO/CCl_4 - 1/3) δ 175.4, 68.6, 45.5, 39.3, 33.9, 29.4, 29.1, 28.9, 22.0, 13.6.

2-(2-Hydroxy-2-methylpropyl)-5-methylhexanehydrazide (5k). Yield 80%, m.p. 158-159 °C. Found, %: C 59.40; H 10.90; N 13.90. $C_{10}H_{22}N_2O_2$. Calculated, %: C 59.37; H 10.96; N 13.85.

^1H NMR (300 MHz, DMSO/CCl_4 - 1/3) δ 8.70 (br.s, 1H, NHC=O), 3.88 (br.s, 3H, NH_2, OH), 3.60 – 3.46 (m, 1H, CHO), 2.16 – 2.02 (m, 1H, CHC=O), 1.70 – 1.56 (m, $1H^a$, CH_2), 1.56 – 1.39 (m, 2H, CH_2), 1.39 – 1.21 (m, 2H, CH_2), 1.17 – 0.96 (m, 3H, CH_3; $1H^b$, CH_2; 1H, $\underline{CH}(CH_3)_2$), 0.87 (dd, J = 6.6, 1.9 Hz, 6H, $CH(\underline{CH_3})_2$).

^{13}C NMR (75 MHz, DMSO/CCl_4 - 1/3) δ 174.6, 64.1, 42.1, 40.9, 35.9, 30.0, 27.5, 23.3, 22.4, 22.1.

2-(2-hydroxypropyl)hexanehydrazide (5l). Yield 67%, m.p. 106-107 °C. R_f 0.61 (ethanol: benz.:hexane-1:1:3,). Found, %: C 57.45; H 10.75; N 14.90. $C_9H_{20}N_2O_2$. Calculated, %: C 57.42; H 10.71; N, 14.88.

4-hydroxy-2-isobutylpentanehydrazide (5m). Yield 74%, m.p. 137-138 °C. R_f 0.65 (ethanol: benz.:hexane-1:1:3,).Found, %: C 57.39; H 10.68; N 14.85. $C_9H_{20}N_2O_2$. Calculated, %: C 57.42; H 10.71; N, 14.88.

2-(2-hydroxypropyl)-5-methylhexanehydrazide(5n). Yield 80%, m.p. 158-159 °C. R_f 0.66 (ethanol: benz.:hexane-1:1:3,).Found, %: C 59.40; H 10.95; N 13.80. $C_{10}H_{22}N_2O_2$. Calculated, %: C 59.37; H 10.96; N 13.85.

2-(2-hydroxy-2-methylpropyl)-5-methylhexanehydrazide (5o). Yield 85%, m.p. 106-108 °C. R_f 0.62 (ethanol: benz.:hexane-1:1:3,). Found, %: C 61.10; H 11.15; N 13.00. $C_{11}H_{24}N_2O_2$. Calculated, %: C 61.08; H 11.18; N 12.95.

^1H NMR (300 MHz, DMSO/CCl_4 - 1/3) δ 8.71 (s, 1H, NHC=O), 3.88 (br.s, 3H, NH_2, OH), 3.53 2,09 (m, 1H, CHO), 1.62 ddd (1H, 13.5, 8.1, 7.2); 1.40 – 1.55 (m, $1H^a$, CH_2), 1.24 – 1.36 (m, 2H, CH_2), 1.08 (2H); 1.05 d (3H, 6,0); 0.87 dd (6H, $2CH_3$, 6.6).

2-(2-hydroxy-2-methylpropyl)hexanehydrazide (5p). Yield 86%, m.p. 105-106 °C. R_f 0.63 (ethanol: benz.:hexane-1:1:3,). Found, %: C 59.40; H 11.00; N 13.80. $C_{10}H_{22}N_2O_2$. Calculated, %: C 59.37; H 10.96; N 13.85.

2-ethyl-4-hydroxy-4-methylpentanehydrazide (5q). Yield 89%, m.p. 75-76 °C. R_f 0.60 (ethanol: benz.:hexane-1:1:3,). Found, %: C 55.18; H 10.45; N 16.10. $C_8H_{18}N_2O_2$. Calculated, %: C 55.15; H 10.41; N 16.08.

2-benzyl-4-hydroxynonanehydrazide (5r). Yield 87%, m.p. 153-154 °C. R_f 0.61 (ethanol: benz.:hexane-1:1:3,). Found, %: C 69.00; H 9.45; N 10.10. $C_{16}H_{26}N_2O_2$. Calculated, %: C 69.03; H 9.41; N 10.06.

2-allyl-4-hydroxynonanehydrazide (s). Yield 87%, m.p. 153-154 °C. R_f 0.64 (ethanol: benz.:hexane-1:1:3,). Found, %: C 63.15; H 10.55; N 12.30. $C_{12}H_{24}N_2O_2$. Calculated, %: C 63.12; H 10.59; N 12.27.

General method for obtaining thiosemicarbazides of 4-hydroxy-2,4-disubstituted butyric acids (6a-t). 0.02 mol of the corresponding acid hydrazide (**2a-f**), 10 ml of ethyl alcohol and 0.02 mol of the corresponding isothiocyanate are placed in a flask. The whole is vigorously stirred until homogenization and left for 3 h. The precipitated crystals are filtered off, washed with ether and dried. Recrystallized from aqueous alcohol (1:1).

The purity and individuality of compounds **5a-t** are determined by TLC on Silufol UV-254 plates.

2-(4-Hydroxy-5-propoxypentanoyl)-N-phenylhydrazinecarbothioamide(6a).
Yield 91%, m.p. 143-144°C. R_f 0,44 (ethanol: benzene: hexane – 1:3:5). Found, %: C 55.40; H 7.15; N 12.85; S 9.90. $C_{15}H_{23}N_3O_3S$. Calculated, %: C 55.36; H 7.12; N 12.91; S 9.85.

^{1}H NMR (300 MHz, DMSO/CCl_4, 1/3) δ 10.42 - 9.76 (m, 1H, NH), 9.68 - 9.19 (m, 2H, NH), 7.49 - 7.41 (m, $2H_{arom}$), 7.40 - 7.32 (m, $2H_{arom}$), 7.12 (ddt, *J* = 8.1, 6.5, 1.3 Hz, $1H_{arom}$), 4.72 (br.s, 1H, OH), 3.52 - 3.41 (m, 1H, $\underline{CH}$OH), 3.38 (dd, *J* = 11.6, 5.2 Hz, $1H_a$, O$\underline{CH_2}$CHOH), 3.14 (dt, *J* = 11.1, 5.1 Hz, $1H_a$, $CH_3CH_2\underline{CH_2}$), 3.06 (dd, *J* = 11.6, 5.2 Hz, $1H_b$, O$\underline{CH_2}$CHOH), 2.88 (dt, *J* = 11.1, 5.1 Hz, $1H_b$, $CH_3CH_2\underline{CH_2}$), 2.32 - 2.13 (m, $1H_a$, $\underline{CH_2}CH_2$CO), 2.12 – 1.88 (m, 2H, $CH_2\underline{CH_2}$CO, $1H_b$, $\underline{CH_2}CH_2$CO), 1.57 (qt, *J* = 7.9, 5.1 Hz, 2H, $CH_3\underline{CH_2}$), 0.80 (t, *J* = 8.0 Hz, 3H, CH_3).

^{13}C NMR (75 MHz, DMSO/CCl_4, 1/3) δ 180.6, 174.0, 141.4, 128.8, 127.3, 123.5, 73.8, 73.6, 70.7, 33.7, 31.6, 23.8, 11.8.

N-Allyl-2-(4-hydroxy-5-propoxypentanoyl)hydrazinecarbothioamide(6b). Yield 77%, m.p. 125-126°C. R_f 0,40 (ethanol: benzene: hexane – 1:3:5). Found, %: C 49.85; H 8.05; N 14.48; S 11.00. $C_{12}H_{23}N_3O_3S$. Calculated, %: C 49.80; H 8.01; N 14.52; S 11.08.

^{1}H NMR (300 MHz, DMSO/CCl_4, 1/3) δ 9.88 - 9.80 (m, 1H, NH<u>NH</u>CS), 9.17 - 9.09 (m, 1H, NHCO), 7.68 (t, J = 4.7 Hz, 1H, <u>NH</u>CH_2), 5.83 (ddt, J = 16.2, 10.8, 5.3 Hz, 1H, <u>CH</u>CH_2NH), 5.11 (ddq, J = 13.6, 2.9, 0.9 Hz, 2H, <u>CH_2</u>CHCH_2NH), 4.72 (br.s, 1H, OH), 4.13 (ddt, J = 5.6, 4.7, 1.0 Hz, 2H, NH<u>CH_2</u>), 3.54 - 3.38 (m, 1H, <u>CH</u>OH), 3.22 (d, J = 5.2 Hz, 2H, O<u>CH_2</u>CHOH), 3.01 (t, J = 5.0 Hz, 2H, CH_3CH_2<u>CH_2</u>), 2.16 - 2.05 (m, 2H, <u>CH_2</u>CH_2CO), 2.04 - 1.94 (m, 2H, <u>CH_2</u>CO), 1.57 (qt, J = 7.9, 5.0 Hz, 2H, CH_3<u>CH_2</u>), 0.80 (t, J = 8.0 Hz, 3H, CH_3).

^{13}C NMR (75 MHz, DMSO/CCl_4, 1/3) δ 182.7, 174.0, 134.8, 115.4, 73.8, 73.6, 70.7, 45.6, 33.7, 31.6, 23.8, 11.8.

2-(2-(2-Hydroxy-3-propoxypropyl)hexanoyl)-N-phenylhydrazinecarbothioamide (6c). Yield 93%, m.p. 153-154°C. R_f 0,43(ethanol: benzene: hexane – 1:3:5). Found, %: C 59.85; H 8.15; N 14.55; S 8.35. $C_{19}H_{31}N_3O_3S$. Calculated, %: C 59.81; H 8.19; N 11.01; S 8.40.

^{1}H NMR (300 MHz, DMSO/CCl_4, 1/3) δ 10.42 - 9.76 (m, 1H, NH), 9.68 - 9.19 (m, 2H, NH), 7.48 - 7.41 (m, $2H_{arom}$), 7.40 - 7.33 (m, $2H_{arom}$), 7.12 (ddt, J = 8.1, 6.5, 1.3 Hz, $1H_{arom}$), 4.72 (br.s, 1H, OH), 3.54 - 3.42 (m, 1H, <u>CH</u>OH), 3.38 (dd, J = 11.6, 5.5 Hz, $1H_a$, O<u>CH_2</u>CHOH), 3.14 (dt, J = 11.1, 5.1 Hz, $1H_a$, CH_3CH_2<u>CH_2</u>O), 3.06 (dd, J = 11.5, 5.5 Hz, $1H_b$, O<u>CH_2</u>CHOH), 2.88 (dt, J = 11.1, 5.1 Hz, $1H_b$, CH_3CH_2<u>CH_2</u>O), 2.30 (tt, J = 8.3, 6.6 Hz, 1H, <u>CH</u>CO), 1.88 (dt, J = 14.7, 8.3 Hz, $1H_a$, CH<u>CH_2</u>CHOH), 1.70 - 1.19 (m, $1H_b$, CH<u>CH_2</u>CHOH, 2H, $CH_3CH_2CH_2$<u>CH_2</u>, 2H, CH_3<u>CH_2</u>CH_2O, 2H, CH_3CH_2<u>CH_2</u>CH_2, 2H, CH_3<u>CH_2</u>CH_2CH_2), 0.91 - 0.74 (m, 6H, CH_3).

^{13}C NMR (75 MHz, DMSO/CCl_4, 1/3) δ 180.0, 177.8, 141.4, 128.8, 127.3, 123.5, 74.1, 73.8, 69.9, 40.5, 36.5, 33.1, 30.3, 23.8, 23.1, 13.8, 11.8.

N-Allyl-2-(2-(2-hydroxy-3-propoxypropyl)hexanoyl)hydrazinecarbothioamide (6d). Yield 89%, m.p. 165-166°C. R_f 0,45(этанол:бензол:гексан – 1:3:5). Found, %: C 55.70; H 9,10; N 12.20; S 9.35. $C_{16}H_{31}N_3O_3S$. Calculated, %: C 55.62; H 9.04; N 12.16; S 9.28.

1H NMR (300 MHz, DMSO) δ 10.31 – 9.65 (m, 1H, NH), 9.61 – 9.12 (m, 2H, NH), 5.78 (ddt, J = 16.5, 10.5, 6.3 Hz, 1H, =CH), 5.19 – 5.12 (m, 1Ha, $=CH_2$), 5.11 (m, 1Hb, $=CH_2$), 4.75 (br.s, 1H, OH), 4.40 (dd, J = 30.4, 6.2 Hz, 2H, NCH_2), 3.56 (m, 1H, OCH), 3.42 (dd, J = 12.3, 5.6 Hz, 1Ha, OCH_2), 3.31 (t, J = 7.6 Hz, 2H, OCH_2), 3.24 (dd, J = 12.3, 5.7 Hz, 1Hb, OCH_2), 2.32 (m, 1H, CHC=O), 1.63 – 1.47 (m, 5H, CH_2), 1.45 – 1.27 (m, 5H, CH_2), 1.02 – 0.93 (m, 6H, CH_3).

13C NMR (75 MHz, DMSO/CCl4 - 1/3) δ 181.9, 175.1, 134.8, 114.8, 74.9, 72.2, 67.4, 45.6, 40.3, 36.7, 31.5, 29.0, 22.4, 22.1, 13.6, 10.3.

2-(5-Butoxy-4-hydroxypentanoyl)-N-phenylhydrazinecarbothioamide(6e). Yield 87%, m.p. 158-159°C. R_f 0,53(ethanol: benzene: hexane – 1:3:5). Found, %: C 56.65; H 7.45; N 12.45; S 9.50. $C_{16}H_{25}N_3O_3S$. Calculated, %: C 56.61; H 7.42; N 12.38; S 9.45.

^1H NMR (300 MHz, DMSO/CCl_4, 1/3) δ 10.42 - 9.76 (m, 1H, NH), 9.68 - 9.19 (m, 2H, NH), 7.47 - 7.41 (m, $2H_{arom}$), 7.40 - 7.32 (m, $2H_{arom}$), 7.12 (ddt, *J* = 8.1, 6.5, 1.3 Hz, $1H_{arom}$), 4.72 (br.s, 1H, OH), 3.53 - 3.41 (m, 1H, <u>CH</u>OH), 3.22 (ddd, *J* = 96.5, 11.6, 5.3 Hz, 2H, O<u>CH</u>$_2$CHOH), 2.95 (td, *J* = 6.7, 0.9 Hz, 2H, $CH_3CH_2CH_2\underline{CH_2}$), 2.22 (m, $1H_a$, $\underline{CH_2}CH_2CO$), 2.01 (m, 2H, $\underline{CH_2}CO$, $1H_b$, $\underline{CH_2}CH_2CO$), 1.57 (pd, *J* = 6.7, 0.6 Hz, 2H, $CH_3CH_2\underline{CH_2}$), 1.36 (m, 2H, $CH_3\underline{CH_2}$), 0.87 (t, *J* = 6.9 Hz, 3H, CH_3).

^{13}C NMR (75 MHz, DMSO/CCl_4, 1/3) δ 180.6, 174.0, 141.4, 128.8, 127.3, 123.5, 73.6, 71.2, 70.7, 33.7, 31.6, 31.4, 17.7, 13.4.

N-Allyl-2-(5-butoxy-4-hydroxypentanoyl)hydrazinecarbothioamide(6f). Yield 92 %, m.p. 124-125°C. R_f 0,46 (ethanol: benzene: hexane – 1:3:5). Found, %: C 51.50; H 8.25; N 13.90; S 10.60. $C_{13}H_{25}N_3O_3S$. Calculated, %: C 51.46; H 8.30; N 13.85; S 10.57.

^1H NMR (300 MHz, DMSO/CCl_4, 1/3) δ 9.88 - 9.80 (m, 1H, NH<u>NH</u>CS), 9.17 - 9.09 (m, 1H, NHCO), 7.68 (t, *J* = 4.7 Hz, 1H, $CH_2\underline{NH}$), 5.83 (ddt, *J* = 16.2, 10.8, 5.3 Hz, 1H, <u>CH</u>CH_2NH), 5.11 (ddq, *J* = 13.6, 2.9, 0.9 Hz, 2H, $\underline{CH_2}CHCH_2NH$),4.72 (br.s, 1H, OH), 4.13 (ddt, *J* = 5.6, 4.7, 1.0 Hz, 2H, $\underline{CH_2}NH$), 3.54 - 3.38 (m, 1H, <u>CH</u>OH), 3.22 (d, *J* = 5.2 Hz, 2H, O<u>CH</u>$_2$CHOH), 2.95 (t, *J* = 6.7 Hz, 2H, $CH_3CH_2CH_2CH_2$), 2.17 - 2.05 (m, 2H, $\underline{CH_2}CH_2CO$), 2.04 - 1.95 (m, 2H, CH_2CO), 1.57 (pd, *J* = 6.7, 0.6 Hz, 2H, $CH_3CH_2\underline{CH_2}$), 1.45 - 1.27 (m, 2H, $CH_3\underline{CH_2}$), 0.87 (t, *J* = 6.9 Hz, 3H, CH_3).

^{13}C NMR (75 MHz, DMSO/CCl_4, 1/3) δ 182.7, 174.0, 134.8, 115.4, 73.6, 71.2, 70.7, 45.6, 33.7, 31.6, 31.4, 17.7, 13.4.

2-(4-Hydroxy-4-(isopentyloxy)butanoyl)-N-phenylhydrazinecarbothioamide(6g). Yield 93%, m.p.131-132°C. R_f 0,43(ethanol: benzene: hexane – 1:3:5). Found, %: C 56.55; H 7.45; N 12.45; S 9.50. $C_{16}H_{25}N_3O_3S$. Calculated, %: C 56.61; H 7.42; N 12.38; S 9.45.

^1H NMR (300 MHz,DMSO/CCl_4, 1/3) δ 10.42 - 9.76 (m, 1H, NH), 9.68 - 9.19 (m, 2H, NH), 7.48 - 7.41 (m, $2H_{arom}$), 7.40 - 7.32 (m, $2H_{arom}$), 7.18 - 7.06 (m, $1H_{arom}$), 4.78 (br.s, 1H, OH), 4.15 (dtd, J = 5.3, 4.3, 1.2 Hz, 1H, $\underline{CH}OH$), 3.01 (ddt, J = 78.8, 11.1, 7.0 Hz, 2H, $CH_3CHCH_2\underline{CH_2}$), 2.14 - 1.90 (m, 2H, CH_2CO, $1H_a$, $\underline{CH_2}CH_2CO$), 1.89 - 1.67 (m, 1H, $CH_3\underline{CH}$, $1H_b$, $\underline{CH_2}CH_2CO$), 1.48 (qd, J = 6.9, 4.5 Hz, 2H, $CH_3CH\underline{CH_2}$), 0.97 (d, J = 6.6 Hz, 6H, CH_3).

^{13}C NMR (75 MHz, DMSO/CCl_4, 1/3) δ 180.6, 174.0, 141.4, 128.8, 127.3, 123.5, 99.7, 67.0, 39.0, 30.1, 29.9, 25.8, 22.9.

N-Allyl-2-(4-hydroxy-4-(isopentyloxy)butanoyl)hydrazinecarbothioamide(6h). Yield 84%, m.p. 102-103°C. R_f 0,54(ethanol: benzene: hexane – 1:3:5). Found, %: C 51.50; H 8.35; N 13.90; S 10.50. $C_{13}H_{25}N_3O_3S$. Calculated, %: C 51.46; H 8.30; N 13.85; S 10.57.

^1H NMR (300 MHz, MHz,DMSO/CCl_4, 1/3) δ 9.88 - 9.80 (m, 1H, NH$\underline{NH}$CS), 9.17 - 9.09 (d, m, 1H, NHCO), 7.68 (t, J = 4.7 Hz, 1H, $CH_2\underline{NH}$), 5.83 (ddt, J = 16.2, 10.7, 5.3 Hz, 1H, $\underline{CH}CH_2NH$), 5.11 (ddq, J = 13.5, 2.9, 0.9 Hz, 2H, $\underline{CH_2}CHCH_2NH$), 4.78 (br.s, 1H, OH), 4.15 (td, J = 5.3, 4.3 Hz, 1H, $\underline{CH}OH$), 3.63 (ddt, J = 5.6, 4.7, 1.0 Hz, 2H, $\underline{CH_2}NH$), 3.01 (t, J = 6.9 Hz, 2H, $CH_3CHCH_2\underline{CH_2}$), 2.08 - 1.95 (m, 2H, CH_2CO), 1.94 - 1.71 (m, 2H, $\underline{CH_2}CH_2CO$, 1H, $CH_3\underline{CH}$), 1.48 (q, J = 6.9 Hz, 2H, $CH_3CH\underline{CH_2}$), 0.97 (d, J = 6.6 Hz, 6H, CH_3).

^{13}C NMR (75 MHz, DMSO/CCl_4, 1/3) δ 182.7, 174.0, 134.8, 115.4, 99.7, 67.0, 45.6, 39.0, 30.1, 29.9, 25.8, 22.9.

2-(2-Benzyl-4-hydroxy-4-(isopentyloxy)butanoyl)-N-phenylhydrazinecarbothioamide(6i). Yield 92%, m.p. 153-154°C. R_f 0,54(ethanol: benzene: hexane – 1:3:8). Found, %: C 64.35; H 7.35; N 9.75; S 7.50. $C_{23}H_{31}N_3O_3S$. Calculated, %: C 64.31; H 7.27; N 9.78; S 7.46.

^1H NMR (300 MHz, MHz, DMSO/CCl_4, 1/3) δ 10.42 - 9.76 (m, 1H, NH), 9.68 - 9.19 (m, 2H, NH), 7.47 - 7.41 (m, $2H_{arom}$), 7.40 - 7.33 (m, $2H_{arom}$), 7.33 - 7.24 (m, $4H_{arom}$), 7.22 - 7.06 (m, $2H_{arom}$), 4.78 (br.s, 1H, OH), 4.15 (td, J = 6.4, 4.5 Hz, 1H, $\underline{CH}OH$), 3.51 (ddt, J = 78.8, 11.1, 7.0 Hz, 2H, $CH_3CHCH_2\underline{CH_2}$), 3.06 (ddt, J = 14.2, 7.2, 1.1 Hz, $1H_a$,

$C_6H_5\underline{CH_2}$), 2.72 (tt, J = 8.5, 7.2 Hz, 1H, CHCO), 2.56 (ddt, J = 14.2, 7.3, 1.1 Hz, $1H_b$, $C_6H_5\underline{CH_2}$), 2.08 (dddd, J = 85.4, 15.2, 8.6, 6.5 Hz, 2H, $CH\underline{CH_2}CHOH$), 1.88 - 1.71 (m, 1H, $CH_3\underline{CH}$), 1.48 (qd, J = 6.8, 4.5 Hz, 2H, $CH_3CH\underline{CH_2}$), 0.97 (d, J = 6.6 Hz, 6H, CH_3).

^{13}C NMR (75 MHz, DMSO/CCl_4, 1/3) δ 180.0, 172.9, 141.4, 139.2, 129.1, 128.8, 128.4, 127.3, 127.0, 123.5, 100.7, 67.0, 40.9, 39.0, 38.2, 36.2, 25.8, 22.9.

2-(2-Benzyl-4-hydroxynonanoyl)-N-phenylhydrazinecarbothioamide (6j). Yield 81%, m.p. 158-159°C. R_f 0,45 (ethanol: benzene: hexane – 1:3:8). Found, %: C 66.85; H 7.50; N 10.20; S 7.70. $C_{23}H_{31}N^3O_2S$. Calculated, %: C 66.79; H 7.56; N 10.16; S 7.75.

^1H NMR (300 MHz, DMSO/CCl_4, 1/3) δ 10.42 - 9.76 (m, 1H, NH), 9.68 - 9.19 (m, 2H, NH), 7.48 - 7.41 (m, $2H_{arom}$), 7.40 - 7.34 (m, $2H_{arom}$), 7.33 - 7.24 (m, $4H_{arom}$), 7.23 - 7.07 (m, $2H_{arom}$), 3.88 (br.s, 1H, OH), 3.11 (ttd, J = 7.7, 6.7, 5.8 Hz, 1H, $\underline{CH}OH$), 2.56 (ddt, J = 14.1, 7.3, 1.0 Hz, $1H_a$, $C_6H_5\underline{CH_2}$), 2.22 (tt, J = 8.3, 7.3 Hz, 1H, CHCO), 2.06 (ddt, J = 14.1, 7.3, 1.0 Hz, $1H_b$, $C_6H_5\underline{CH_2}$), 1.75 (dddd, J = 79.4, 14.7, 8.4, 7.6 Hz, 2H, $CH\underline{CH_2}CHOH$), 1.56 - 1.41 (m, $1H_a$, $CH_3CH_2CH_2CH_2\underline{CH_2}$), 1.40 - 1.15 (m, 2H, $CH_3CH_2CH_2\underline{CH_2}$, 2H, $CH_3CH_2\underline{CH_2}$, $1H_b$, $CH_3CH_2CH_2CH_2\underline{CH_2}$, 2H, $CH_3\underline{CH_2}$), 0.91 - 0.78 (m, 3H, CH_3).

^{13}C NMR (75 MHz, DMSO/CCl_4, 1/3) δ 180.0, 172.9, 141.4, 139.2, 129.1, 128.8, 128.4, 127.3, 127.0, 123.5, 67.6, 43.4, 39.3, 38.0, 36.2, 31.8, 25.5, 22.8, 14.1.

2-(2-(2-hydroxy-2-methylpropyl)hexanoyl)-N-phenylhydrazine-1-carbothioamide(6k). Yield 76%, m.p. 135-137°C. R_f 0,56 (ethanol: benzene: hexane – 1:3:8). Found, %: C 60.50; H 8.20; N 12.50; S 9.60. $C_{17}H_{27}N_3O_2S$. Calculated, %: C 60.53; H 8.01; N 12.46; S 9.50.

^1H NMR (300 MHz, DMSO/CCl_4, 1/3, 0.86 d & 0.90 d (6H, $2CH_3$, 6.6): 1.04 m (1H, NH), 1.31 dt (1H, 14.1, 2.9); 1.44-1.59 m (2H) ; 1.70 dt (1H, 14.1, 10.2) ; 2.31 m (1H) ; 3.68 M (1H, OCH) ; 3.97 dtt (1H, CH_2); 4.17 dtt (1H, 15,6, 5.7, 1.6); 4.53 br (1H, OH) ; 5.04 ddt (1H, $=CH_2$, 10.2, 1.6, 1.6); 5.16 ddt (1H, $=CH_2$, 17.1, 1.7, 1.7); 5.84 ddt (1H, $=CH_2$, 17.1, 10.2, 5.4).

2-(2-(2-hydroxy-2-methylpropyl)-5-methylhexanoyl)-N-phenylhydrazine-1-carbothioamide (6l). Yield 75%, m.p. 138-140°C. R_f 0,57 (ethanol: benzene: hexane – 1:3:8). Found, %: C 61.50; H 8.40; N 12.05; S 9.20. $C_{18}H_{30}N_3O_2S$. Calculated, %: C 61.54; H 8.51; N 11.93; S 9.09.

N-allyl-2-(2-(2-hydroxy-2-methylpropyl)-5-methylhexanoyl)hydrazine-1-carbothioamide (6m). Yield 81%, m.p. 135-136°C. R_f 0,55 (ethanol: benzene: hexane – 1:3:8). Found, %: C 57.25; H 9.05; N 13.50; S 10.10. $C_{15}H_{29}N_3O_2S$. Calculated, %: C 57.25; H 9.15; N 13.40; S 10.16.

^{1}H NMR (300 MHz, DMSO/CCl_4, 1/3, 0.90 t (3H, CH_3, 6.9): 1.11 d (3H, CH_3, 6.3), 1.15-1.40 m (6H); 1.51 m (1H) ; 1.72 dt (1H, 13.8, 10.2); 2.20 m (1H); 3.66 m (1H, OCH) ; 3.99 dtt (1H, CH_2,15.6, 5.7, 1.6); 4.17 dtt (1H, 15,6, 5.7, 1.6); 4.49 d (1H, OH, 6.6); 5.04 ddt (1H, =CH_2, 10.2, 1.6, 1.6); 5.16 ddt (1H, =CH_2, 17.1, 1.6, 1.6); 5.84 ddt (1H, =CH, 17.1, 10.2, 5.4); 7.95 t (1H, NH, 5.7); 9.01 s (1H, NH) & 9.61 s (1H, NH).

2-(2-ethyl-4-hydroxy-4-methylpentanoyl)-N-phenylhydrazine-1-carbothioamide(6n). Yield 90%, m.p. 198-199°C. R_f 0,63 (ethanol: benzene: hexane – 1:3:8). Found, %: C 58.30; H 7.46; N 13.62; S 10.30. $C_{15}H_{23}N_3O_2S$. Calculated, %: C 58.25; H 7.44; N 13.59; S 10.36.

N-allyl-2-(2-ethyl-4-hydroxy-4-methylpentanoyl)hydrazine-1-carbothioamide(6o). Yield 82%, m.p. 142-143°C. R_f 0,62 (ethanol: benzene: hexane – 1:3:8). Found, %: C 52.30; H 8.46; N 15.40; S 11.75. $C_{12}H_{23}N_3O_2S$. Calculated, %: C C 52.75; H 8.48; N 15.37; S 11.73.

N-allyl-2-(4-hydroxy-2-isobutylpentanoyl)hydrazine-1-carbothioamide(6p). Yield 85%, m.p. 145-146°C. R_f 0,60 (ethanol: benzene: hexane – 1:3:8). Found, %: C 54.30; H 8.78; N 14.65; S 11.20. $C_{13}H_{25}N_3O_2S$. Calculated, %: C 54.33; H 8.77; N 14.62; S 11.15.

^{1}H NMR (300 MHz, DMSO/CCl_4, 1/3, 0.90 t (3H, CH_3, 6.9): 1.11 d (3H, CH_3, 6.3), 1.15-1.40 m (6H); 1.51 m (1H) ; 1.72 dt (1H, 13.8, 10.2); 2.20 m (1H); 3.66 m (1H, OCH) ; 3.99 dtt (1H, CH_2,15.6, 5.7, 1.6); 4.17 dtt (1H, 15,6, 5.7, 1.6); 4.49 d (1H, OH, 6.6); 5.04 ddt (1H, =CH_2, 10.2, 1.6, 1.6); 5.16 ddt (1H, =CH_2, 17.1, 1.6, 1.6); 5.84 ddt (1H, =CH, 17.1, 10.2, 5.4); 7.95 t (1H, NH, 5.7); 9.01 s (1H, NH) & 9.61 s (1H, NH).

N-allyl-2-(2-(2-hydroxypropyl)hexanoyl)hydrazine-1-carbothioamide(6q). Yield 95%, m.p. 153-154°C. R_f 0,58 (ethanol: benzene: hexane – 1:3:8). Found, %: C 54.35; H 8.80; N 14.63; S 11.18. $C_{13}H_{25}N_3O_2S$. Calculated, %: C 54.33; H 8.77; N 14.62; S 11.15.

^{1}H NMR (300 MHz, DMSO/CCl_4, 1/3, 0.90 t (3H, CH_3, 6.9): 1.11 d (3H, CH_3, 6.3), 1.15-1.40 m (6H); 1.51 m (1H) ; 1.72 dt (1H, 13.8, 10.2); 2.20 m (1H); 3.66 m (1H, OCH) ; 3.99 dtt (1H, CH_2,15.6, 5.7, 1.6); 4.17 dtt (1H, 15,6, 5.7, 1.6); 4.49 d (1H, OH,

6.6); 5.04 ddt (1H, $=CH_2$, 10.2, 1.6, 1.6); 5.16 ddt (1H, $=CH_2$, 17.1, 1.6, 1.6); 5.84 ddt (1H, =CH, 17.1, 10.2, 5.4); 7.95 t (1H, NH, 5.7); 9.01 s (1H, NH) & 9.61 s (1H, NH).

N-allyl-2-(2-(2-hydroxypropyl)-5-methylhexanoyl)hydrazine-1-carbothioamide(5r). Yield 94%, m.p. 152-153°C. R_f 0,58 (ethanol: benzene: hexane – 1:3:8). Found, %: C 55.80; H 9.09; N 14.00; S 11.18. $C_{13}H_{27}N_3O_2S$. Calculated, %: C 55.78; H 9.03; N 13.94 ; S 11.15.

^{1}H NMR (300 MHz, DMSO/CCl_4, 1/3, 0.87 d & 0.88 d (6H, $2CH_3$, 6.6): 1.11 d (3H, CH_3, 6.3), 1.14 m (2H) ; 1.23-1.40 m (2H); 1.41-1.59 m (2H) ; 1.72 dt (1H, 14.1, 10.2); 2.17 m (1H); 3.66 m (1H); 3.99 dtt (1H, CH_2, 15.5, 5.7, 1.8); 4.17 dtt (1H, 15,5, 5.7, 1.8); 4.50 d (1H, OH, 6.6); 5.04 ddt (1H, $=CH_2$, 10.2, 1.7, 1.7); 5.16 ddt (1H, $=CH_2$, 17.1, 1.8, 1.8); 5.84 ddt (1H, =CH, 17.1, 10.2, 5.4); 7.94 t (1H, NH, 5.7); 9.01 s (1H, NH, 5.7) ; 9.61 s (1H, NH).

2-(2-(2-hydroxypropyl)hexanoyl)-N-phenylhydrazine-1-carbothioamide(6s). Yield 81%, m.p. 146-147°C. R_f 0,64 (ethanol: benzene: hexane – 1:3:8). Found, %: C 59.45; H 7.85; N 12.95; S 9.95. $C_{16}H_{25}N_3O_2S$. Calculated, %: C 59.41; H 7.79; N 12.99; S 9.91.

2-(2-(2-hydroxypropyl)-5-methylhexanoyl)-N-phenylhydrazine-1-carbothioamide(6t). Yield 93%, m.p. 156-177°C. R_f 0,60 (ethanol: benzene: hexane – 1:3:8). Found, %: C 60.45; H 8.10; N 12.48; S 9.45. $C_{17}H_{27}N_3O_2S$. Calculated, %: C 60.50; H 8.06; N 12.45; S 9.50.

General method for obtaining 3,4-disubstituted-5-mercapto-4*H*-1,2,4-triazoles (7a-t). 3g (0.075 mol) of sodium hydroxide in the form of a 10% aqueous solution, 0.05 mol of the corresponding **2a-j** are placed in a flask and stirred. The mixture is heated in a boiling water bath for 4 hours, cooled, acidified with a dilute solution of hydrochloric acid to pH 6. The precipitated crystals are filtered off, washed with water and dried. Recrystallized from aqueous alcohol (1:1).

The purity and individuality of compounds **4a-j** are determined by TLC on Silufol UV-254 plates.

3-(3-Hydroxy-4-propoxybutyl)-4-phenyl-1*H*-1,2,4-triazole-5(4*H*)-thione (7a). Yield 90%, m.p. 100-101°C. R_f 0,65 (ethanol: benzene: hexane – 1:3:5). Found, %: C 58.65; H 6.85; N 13.70; S 10.50. $C_{15}H_{21}N_3O_2S$. Calculated, %: C 58.61; H 6.89; N 13.67; S 10.43.

^{1}H NMR (300 MHz, DMSO/CCl_4, 1/3) δ 13.26 (br.s, 1H, NH), 7.61 (ddt, *J* = 7.7, 6.5, 1.3 Hz, $1H_{arom}$), 7.54 - 7.42 (m, $2H_{arom}$), 7.41 - 7.31 (m, $2H_{arom}$), 4.72 (br.s, 1H, OH), 3.52 - 3.42 (m, 1H, $\underline{CH}OH$), 3.38 (dd, *J* = 11.6, 5.1 Hz, $1H_a$, $O\underline{CH_2}CHOH$), 3.14 (dt, *J* = 11.1, 5.1 Hz, $1H_a$, $CH_3CH_2\underline{CH_2}$), 3.06 (dd, *J* = 11.6, 5.1 Hz, $1H_b$, $O\underline{CH_2}CHOH$), 2.88 (dt, *J* = 11.1, 5.1 Hz, $1H_b$, $CH_3CH_2\underline{CH_2}$), 2.16 - 1.98 (m, 2H, $\underline{CH_2}CH_2CHOH$), 1.79 - 1.45 (m, 2H, $CH_2\underline{CH_2}CHOH$, 2H, $CH_3\underline{CH_2}$), 0.80 (t, *J* = 8.0 Hz, 3H, CH_3).

^{13}C NMR (75 MHz, DMSO/CCl_4, 1/3) δ 175.8, 151.7, 140.1, 128.9, 128.1, 127.4, 73.8, 73.6, 70.7, 34.9, 25.1, 23.8, 11.8.

4-Allyl-3-(3-hydroxy-4-propoxybutyl)-1*H*-1,2,4-triazole-5(4*H*)-thione (7b). Yield 92%, m.p. 102-103°C. R_f 0,71 (ethanol: benzene: hexane – 1:3:5). Found, %: C 53.15; H 7.85; N 15.55; S 11.75. $C_{12}H_{21}N_3O_2S$. Calculated, %: C 53.11; H 7.80; N 15.48; S 11.82.

^{1}H NMR (300 MHz, DMSO/CCl_4, 1/3) δ 13.48 (br.s, 1H, NH), 5.66 - 5.46 (m, 1H, $CH_2\underline{CH}CH_2N$), 4.72 (br.s, 1H, OH), 4.67 - 4.58 (m, 2H, $\underline{CH_2}CHCH_2N$), 4.50 (dt, *J* = 6.8, 1.0 Hz, 2H, $CH_2CH\underline{CH_2}N$), 3.46 (tdt, *J* = 7.1, 5.9, 4.9 Hz, 1H, $\underline{CH}OH$), 3.22 (d, *J* = 5.1 Hz, 2H, $O\underline{CH_2}CHOH$), 3.01 (d, *J* = 10.1 Hz, 2H, $CH_3CH_2\underline{CH_2}$), 2.03 (t, *J* = 9.1 Hz, 2H, $\underline{CH_2}CH_2CHOH$), 1.69 - 1.48 (m, 2H, $CH_2\underline{CH_2}CHOH$, 2H, $CH_3\underline{CH_2}$), 0.80 (t, *J* = 8.0 Hz, 3H, CH_3).

^{13}C NMR (75 MHz, DMSO/CCl_4, 1/3) δ 165.4, 147.5, 133.1, 117.6, 73.8, 73.6, 70.7, 45.1, 34.9, 24.1, 23.8, 11.8.

3-(2-Hydroxy-1-propoxyoctan-4-yl)-4-phenyl-1*H*-1,2,4-triazole-5(4*H*)-thione (7c). Yield 91%, m.p. 103-105°C. R_f 0,69 (ethanol: benzene: hexane – 1:3:5). Found, %: C 62.78; H 8.04; N 11.56; S 8.82. $C_{19}H_{29}N_3O_2S$. Calculated , %: C 62.78; H 8.04; N 11.56; S 8.82.

^{1}H NMR (300 MHz, DMSO/CCl_4, 1/3) δ 13.01 (br.s, 1H, NH), 7.61 (ddt, *J* = 7.7, 6.5, 1.3 Hz, $1H_{arom}$), 7.54 - 7.42 (m, $2H_{arom}$), 7.41 - 7.31 (m, $2H_{arom}$), 4.72 (br.s, 1H, OH), 3.53 - 3.42 (m, 1H, $\underline{CH}OH$), 3.38 (dd, *J* = 11.5, 5.3 Hz, $1H_a$, $O\underline{CH_2}CH$), 3.14 (dt, *J* = 11.1, 5.1 Hz, $1H_a$, $CH_3CH_2\underline{CH_2}O$), 3.06 (dd, *J* = 11.6, 5.3 Hz, $1H_b$, $O\underline{CH_2}CH$), 2.88 (dt, *J* = 11.1, 5.1 Hz, $1H_b$, $CH_3CH_2\underline{CH_2}O$), 2.08 (tt, *J* = 8.7, 7.0 Hz, 1H, $\underline{CH}CH_2CHOH$), 1.90 - 1.71 (m, $1H_a$, $CH\underline{CH_2}CHOH$, $1H_a$, $CH_3CH_2CH_2\underline{CH_2}$), 1.65 - 1.51 (m, $1H_b$, $CH\underline{CH_2}CHOH$, 2H, $CH_3\underline{CH_2}CH_2O$), 1.50 - 1.38 (m, $1H_b$, $CH_3CH_2CH_2\underline{CH_2}$), 1.37 - 1.18 (m, 2H, $CH_3CH_2\underline{CH_2}CH_2$), 1.17 - 0.95 (m, 2H, $CH_3\underline{CH_2}CH_2CH_2$), 0.86 - 0.74 (m, 3H, $\underline{CH_3}CH_2CH_2O$, 3H, $\underline{CH_3}CH_2CH_2CH_2$).

^{13}C NMR (75 MHz, DMSO/CCl_4, 1/3) δ 173.7, 152.8, 135.0, 129.3, 128.1, 127.9, 74.1, 73.8, 69.9, 38.9, 36.2, 33.8, 30.3, 23.8, 23.1, 13.8, 11.8.

4-Allyl-3-(2-hydroxy-1-propoxyoctan-4-yl)-1*H*-1,2,4-triazole-5(4*H*)-thione (7d). Yield 86%, m.p. 84-85°C. R_f 0,68 (ethanol: benzene: hexane – 1:3:5). Found, %: C 58.65; H 9.00; N 12.85; S 9.85. $C_{16}H_{29}N_3O_2S$. Calculated, %: C 58.68; H 8.93; N 12.83; S 9.79.

^1H NMR (300 MHz, DMSO/CCl_4, 1/3) δ 13.49 (br.s, 1H, NH), 5.66 - 5.46 (m, 1H, $CH_2\underline{CH}CH_2N$), 4.72 (br.s, 1H, OH), 4.67 - 4.54 (m, 2H, $\underline{CH_2}CHCH_2N$), 3.95 (dt, J = 7.1, 1.0 Hz, 2H, $CH_2CH\underline{CH_2}N$), 3.46 (tdt, J = 8.2, 6.1, 5.2 Hz, 1H, $\underline{CH}OH$), 3.52 (d, J = 5.3 Hz, 2H, $O\underline{CH_2}CH$), 3.01 (d, J = 10.1 Hz, 2H, $CH_3CH_2\underline{CH_2}O$), 2.08 (tt, J = 8.9, 7.1 Hz, 1H, $\underline{CH}CH_2CHOH$), 1.75 - 1.48 (m, 2H, $CH\underline{CH_2}CHOH$, 2H, $CH_3CH_2CH_2\underline{CH_2}$, 2H, $CH_3\underline{CH_2}CH_2O$), 1.36 - 1.83 (m, 2H, $CH_3\underline{CH_2}CH_2CH_2$), 1.16 - 0.99 (m, 2H, $CH_3CH_2\underline{CH_2}CH_2$), 0.86 - 0.74 (m, 3H, $\underline{CH_3}CH_2CH_2O$, 3H, $\underline{CH_3}CH_2CH_2CH_2$).

^{13}C NMR (75 MHz, DMSO/CCl_4, 1/3) δ 133.1, 117.6, 74.1, 73.8, 69.9, 45.3, 37.1, 36.2, 33.8, 30.3, 23.8, 23.1, 13.8, 11.8.

3-(4-Butoxy-3-hydroxybutyl)-4-phenyl-1*H*-1,2,4-triazole-5(4*H*)-thione (7e). Yield 89%, m.p. 109-110°C. R_f 0,67 (ethanol: benzene: hexane – 1:3:5). Found, %: C 59.75; H 7.25; N 13.15; S 9.95. $C_{16}H_{23}N_3O_2S$. Calculated, %: C 59.78; H 7.21; N 13.07; S 9.98.

^1H NMR (300 MHz, DMSO/CCl_4, 1/3) δ 13.78 (br.s, 1H, NH), 7.61 (ddt, J = 7.7, 6.5, 1.3 Hz, $1H_{arom}$), 7.54 - 7.42 (m, $2H_{arom}$), 7.41 - 7.31 (m, $2H_{arom}$), 4.72 (br.s, 1H, OH), 3.53 - 3.42 (m, 1H, CHOH), 3.22 (ddd, J = 96.5, 11.6, 5.1 Hz, 2H, $O\underline{CH_2}CH$), 2.95 (td, J = 6.7, 0.9 Hz, 2H, $\underline{CH_2}OCH_2CH$), 2.16 - 1.90 (m, 2H, CH_2C), 1.79 -1.45 (m, 4H, $CH_3CH_2\underline{CH_2}$, $\underline{CH_2}CH_2C$), 1.45 - 1.27 (m, 2H, $CH_3\underline{CH_2}$), 0.87 (t, J = 6.9 Hz, 3H, CH_3).

^{13}C NMR (75 MHz, DMSO/CCl_4, 1/3) δ 175.8, 151.7, 140.0, 128.9, 128.1, 127.4, 73.6, 71.2, 70.7, 34.9, 31.4, 25.1, 17.7, 13.4.

4-Allyl-3-(4-butoxy-3-hydroxybutyl)-1*H*-1,2,4-triazole-5(4*H*)-thione (7f). Yield 87%, m.p. 116-117°C. R_f 0,70 (ethanol: benzene: hexane – 1:3:5). Found, %: C 54.75; H 8.05; N 14.75; S 11.20. $C_{13}H_{23}N_3O_2S$. Calculated, %: C 54.71; H 8.12; N 14.72; S 11.23.

^1H NMR (300 MHz, DMSO/CCl_4, 1/3) δ 12.81 (br.s, 1H, NH), 5.66 - 5.46 (m, 1H, $CH_2\underline{CH}CH_2N$), 4.72 (br.s, 1H, OH), 4.67 - 4.56 (m, 2H, $\underline{CH_2}CHCH_2N$), 4.51 (dt, J =

6.8, 1.0 Hz, 2H, $CH_2CH\underline{CH_2}N$), 3.46 (tdt, *J* = 7.1, 5.9, 5.0 Hz, 1H, <u>CH</u>OH), 3.22 (d, *J* = 5.1 Hz, 2H, $O\underline{CH_2}CHOH$), 2.95 (d, *J* = 13.5 Hz, 2H, $CH_3CH_2CH_2\underline{CH_2}$), 2.03 (t, *J* = 9.1 Hz, 2H, $\underline{CH_2}CH_2CHOH$), 1.60 (m, 2H, $CH_2\underline{CH_2}CHOH$, 2H, $CH_3CH_2\underline{CH_2}$), 1.36 (m, 2H, $CH_3\underline{CH_2}$), 0.87 (t, *J* = 6.9 Hz, 3H, CH_3).

^{13}C NMR (75 MHz, DMSO/CCl_4, 1/3) δ 165.4, 147.5, 133.1, 117.6, 73.6, 71.2, 70.7, 45.1, 34.9, 31.4, 24.1, 17.7, 13.4.

3-(3-Hydroxy-4-(isopentyloxy)butyl)-4-phenyl-1*H*-1,2,4-triazole-5(*4H*)-thione (7g). Yield 91%, m.p.121-122°C. R_f 0,57 (ethanol: benzene: hexane – 1:3:5). Found, %: C 60.90; H 7.55; N 12.50; S 9.60. $C_{17}H_{25}N_3O_2S$. Calculated, %: C 60.87; H 7.51; N 12.53; S 9.56.

^1H NMR (300 MHz, DMSO/CCl_4, 1/3) δ 13.00 (br.s, 1H, NH), 7.61 (ddt, *J* = 7.8, 6.5, 1.4 Hz, $1H_{arom}$), 7.54 - 7.42 (m, $2H_{arom}$), 7.41 - 7.31 (m, $2H_{arom}$), 4.72 (br.s, 1H, OH), 3.52 - 3.42 (m, 1H, <u>CH</u>OH), 3.22 (ddd, *J* = 96.5, 11.6, 5.1 Hz, 2H, $O\underline{CH_2}CHOH$), 2.90 (td, *J* = 6.9, 0.9 Hz, 2H, $CH_3CHCH_2\underline{CH_2}$), 2.16 - 1.90 (m, 2H, $\underline{CH_2}CH_2CHOH$), 1.89 - 1.76 (m, 1H, $CH_3\underline{CH}$), 1.75 - 1.47 (m, 2H, $CH_2\underline{CH_2}CHOH$), 1.41 (q, *J* = 6.8 Hz, 2H, $CH_3CH\underline{CH_2}$), 0.97 (d, *J* = 6.6 Hz, 6H, CH_3).

^{13}C NMR (75 MHz, DMSO/CCl_4, 1/3) δ 175.8, 151.7, 140.1, 128.9, 128.1, 127.4, 73.6, 70.7, 69.9, 38.6, 34.9, 25.1, 25.0, 22.6.

4-Allyl-3-(3-hydroxy-4-(isopentyloxy)butyl)-1*H*-1,2,4-triazole-5(*4H*)-thione (7h). Yield 89%, m.p. 97-98°C. R_f 0,76 (ethanol: benzene: hexane – 1:3:5). Found, %: C 56.20; H 8.45; N 14.00; S 10.75. $C_{14}H_{25}N_3O_2S$. Calculated, %: C 56.16; H 8.42; N 14.03; S 10.71.

^1H NMR (300 MHz, DMSO/CCl_4, 1/3) δ 14.00 (br.s, 1H, NH), 5.66 - 5.46 (m, 1H, $CH_2\underline{CH}CH_2N$), 4.72 (br.s, 1H, OH), 4.63 - 4.51 (m, 2H, $\underline{CH_2}CHCH_2N$), 4.45 (dt, *J* = 6.8, 1.0 Hz, 2H, $CH_2CH\underline{CH_2}N$), 3.46 (tdt, *J* = 7.1, 5.9, 4.9 Hz, 1H, <u>CH</u>OH), 3.22 (d, *J* = 5.1 Hz, 2H, $O\underline{CH_2}CHOH$), 2.90 (d, *J* = 13.7 Hz, 2H, $CH_3CHCH_2\underline{CH_2}$), 2.03 (t, *J* = 9.1 Hz, 2H, $\underline{CH_2}CH_2CHOH$), 1.81 (dp, *J* = 13.2, 6.6 Hz, 1H, $CH_3\underline{CH}$), 1.62 (td, *J* = 9.1, 7.1 Hz, 2H, $CH_2\underline{CH_2}CHOH$), 1.41 (q, *J* = 6.8 Hz, 2H, $CH_3CH\underline{CH_2}$), 0.97 (d, *J* = 6.6 Hz, 6H, CH_3).

^{13}C NMR (75 MHz, DMSO/CCl_4, 1/3) δ 165.4, 147.5, 133.1, 117.6, 73.6, 70.7, 69.9, 45.1, 38.6, 34.9, 25.0, 24.1, 22.6.

3-(4-Hydroxy-5-(isopentyloxy)-1-phenylpentan-2-yl)-4-phenyl-1*H*-1,2,4-triazole-5(4*H*)-thione (7i). Yield 84%, m.p. 95-96°C. R_f 0,61 (ethanol: benzene: hexane – 1:3:8). Found, %: C 67.70; H 7.30; N 9.90; S 7.55. $C_{24}H_{31}N_3O_2S$. Calculated, %: C 67.73; H 7.34; N 9.87; S 7.53.

^{1}H NMR (300 MHz, DMSO/CCl_4, 1/3) δ 13.66 (br.s, 1H, NH), 7.61 (ddt, *J* = 7.7, 6.5, 1.3 Hz, 1H_{arom}), 7.54 - 7.42 (m, 2H_{arom}), 7.41 - 7.31 (m, 2H_{arom}), 7.29 - 7.10 (m, 5H_{arom}), 4.72 (br.s, 1H, OH), 4.02 - 3.92 (m, 1H, $\underline{CH}$OH), 3.52 (ddd, *J* = 96.5, 11.5, 5.3 Hz, 2H, $O\underline{CH_2}$CHOH), 2.90 (td, *J* = 6.9, 0.9 Hz, 2H, $CH_3CHCH_2\underline{CH_2}$), 2.75 - 2.42 (m, 2H, $C_6H_5\underline{CH_2}$, 1H, $C_6H_5CH_2\underline{CH}$), 1.91 - 1.69 (m, 1H_a, $CH\underline{CH_2}$CHOH, 1H, $CH_3\underline{CH}$), 1.57 (ddd, *J* = 14.8, 8.7, 8.0 Hz, 1H_b, $CH\underline{CH_2}$CHOH), 1.41 (q, *J* = 6.8 Hz, 2H, $CH_3CH\underline{CH_2}$), 0.97 (d, *J* = 6.6 Hz, 6H, CH_3).

^{13}C NMR (75 MHz, DMSO/CCl_4, 1/3) δ 173.7, 148.0, 139.2, 135.0, 129.3, 129.1, 128.4, 128.1, 127.9, 127.0, 74.1, 69.9, 43.7, 38.6, 36.6, 34.4, 25.0, 22.6.

3-(4-hydroxy-1-phenylnonan-2-yl)-4-phenyl-1*H*-1,2,4-triazole-5(4*H*)-thione (7j). Yield 81%, m.p. 161-162°C. R_f 0,60 (ethanol: benzene: hexane – 1:3:3). Found, %: C 69.84; H 7.39; N 10.62; S 8.11. $C_{23}H_{29}N_3OS$. Calculated, %: C 69.84; H 7.39; N 10.62; S 8.11.

^{1}H NMR (300 MHz, DMSO/CCl_4, 1/3) δ 13.81 (br.s, 1H, NH), 7.61 (ddt, *J* = 7.7, 6.5, 1.3 Hz, 1H_{arom}), 7.54 - 7.42 (m, 2H_{arom}), 7.41 - 7.31 (m, 2H_{arom}), 7.30 - 7.12 (m, 5H_{arom}), 3.88 (br.s, 1H, OH), 3.11 (ttd, *J* = 7.6, 6.5, 5.6 Hz, 1H, $\underline{CH}$OH), 2.73 - 2.47 (m, 2H, $C_6H_5\underline{CH_2}$, 1H, $C_6H_5CH_2\underline{CH}$), 1.81 (ddd, *J* = 14.8, 8.8, 7.6 Hz, 1H_a, $CH\underline{CH_2}$CHOH), 1.64 - 1.42 (m, 1H_b, $CH\underline{CH_2}$CHOH, 1H_a, $CH_3CH_2CH_2CH_2\underline{CH_2}$), 1.41 - 1.16 (m, 2H, $CH_3CH_2CH_2\underline{CH_2}$, 2H, $CH_3CH_2\underline{CH_2}$, 1H_b, $CH_3CH_2CH_2CH_2\underline{CH_2}$, 2H, $CH_3\underline{CH_2}$), 0.91 - 0.78 (m, 3H, CH_3).

^{13}C NMR (75 MHz, DMSO/CCl_4, 1/3) δ 173.7, 148.0, 139.2, 135.0, 129.3, 129.1, 128.4, 128.1, 127.9, 127.0, 67.6, 43.7, 38.1, 38.0, 36.6, 31.8, 25.5, 22.7, 14.1.

4-allyl-5-(2-hydroxy-2,7-dimethyloctan-4-yl)-2,4-dihydro-3H-1,2,4-triazole-3-thione (7k). Yield 69%, m.p. 132-133°C. R_f 0,51 (ethanol: benzene: hexane – 1:3:3). Found, %: C 60.65; H 9.10; N 14.15; S 10.75. $C_{15}H_{27}N_3OS$. Calculated, %: C 60.57; H 9.15; N 14.13; S 10.78.

4-allyl-5-(2-hydroxy-2,7-dimethyloctan-4-yl)-2,4-dihydro-3H-1,2,4-triazole-3-thione (7l). Yield 73%, m.pm., 88-89°C. R_f 0,52 (ethanol: benzene: hexane – 1:3:3).

Found, %: C 64.80; H 8.10; N 14.57; S 9.65. $C_{18}H_{27}N_3OS$. Calculated, %: C 64.83; H 8.16; N 12.60; S 9.61.

5-(2-hydroxy-2-methyloctan-4-yl)-4-phenyl-2,4-dihydro-3H-1,2,4-triazole-3-thione (7m). Yield 75%, m.p. 115-116°C. R_f 0,54 (ethanol: benzene: hexane – 1:3:3). Found, %: C 63.85; H 7.95; N 13.20; S 10.10. $C_{17}H_{25}N_3OS$. Calculated, %: C 63.91; H 7.89; N 13.15; S 10.04

4-allyl-5-(5-hydroxy-5-methylhexan-3-yl)-2,4-dihydro-3H-1,2,4-triazole-3-thione (7n). Yield 58%, m.p. 134-135°C. R_f 0,59 (ethanol: benzene: hexane – 1:3:3). Found, %: C 56.45; H 8.30; N 16.40; S 12.50. $C_{12}H_{21}N_3OS$. Calculated, %: C 56.44; H 8.29; N 16.45; S 12.55.

5-(5-hydroxy-5-methylhexan-3-yl)-4-phenyl-2,4-dihydro-3H-1,2,4-triazole-3-thione (7o). Yield 70%, m.p. 141-142°C. R_f 0,57 (ethanol: benzene: hexane – 1:3:3). Found, %: C 61.85; H 7.30; N 14.45; S 10.95. $C_{15}H_{21}N_3OS$. Calculated, %: C 61.82; H 7.26; N 14.42; S 11.00.

4-allyl-5-(2-hydroxy-6-methylheptan-4-yl)-2,4-dihydro-3H-1,2,4-triazole-3-thione (7p). Yield 80%, m.p. 98-99°C. R_f 0,58 (ethanol: benzene: hexane – 1:3:3). Found, %: C 58.00; H 8.65; N 15.65; S 11.95. $C_{13}H_{23}N_3OS$. Calculated, %: C 57.96; H 8.61; N 15.60; S 11.90.

^{1}H NMR (300 MHz, DMSO/CCl_4, 1/3, 0.76 d & 0.82 d (3H, CH_3, 6.9): 0.99 d & 1.04 d(3H, CH_3, 6.0), 1.18 m (2H) ; 1.0-1.2 m (4H); 1.25-1.60 m (3H) ; 1.79 dt (1H); 2.59 m & 2.82 m (1H); 3.56 m (1H, OCH); 4.03 d & 4.14 d (1H, OH, 5.1); 7,3-7.4 m (2H, o-C_6H_5); 7.45-7.60 m (3H, m- & p-C_6H_5, 7.8); 13.48 s & 13.51 s (1H, SH).

4-allyl-5-(2-hydroxyoctan-4-yl)-2,4-dihydro-3H-1,2,4-triazole-3-thione (7q). Yield 73%, 118-119°C. R_f 0,56 (ethanol: benzene: hexane – 1:3:3). Found, %: C 62.95; H 7.65; N 13.80; S 10.55. $C_{16}H_{23}N_3OS$. Calculated, %: C 57.96; H 8.61; N 15.60; S, 11.90

^{1}H NMR (300 MHz, DMSO/CCl_4, 1/3, 0.87 t (3H, CH_3, 7.0); 1.06 d (3H, CH_3, 6.0): 1.10-1.35 m (4H) 1.45-1.80 m (4H); 2.90 m (1H); 3.68 m (1H, OCH); 4.27 d (1H, OH, 5.4); 4.57 ddt (1H, NCH_2, 15.9, 6.3, 1.5); 4.71 ddt (1H, NCH_2, 15.9, 4.8, 1.8); 5.17 dd (1H, =CH_2, 17.0, 1.5); 5.20 dd (1H, =CH_2, 10.5, 1.5); 5.89 dd (1H, 17.0, 10.5, 6.3, 4.8); 13.30 (1H, SH).

4-allyl-5-(2-hydroxy-6-methylheptan-4-yl)-2,4-dihydro-3H-1,2,4-triazole-3-thione (7r). Yield 70%, 98-99°C. R_f 0,58 (ethanol: benzene: hexane – 1:3:3). Found,

%: C 58.00; H 8.65; N 14.80; S 11.25. $C_{13}H_{23}N_3OS$. Calculated, %: C 57.96; H 8.61; N 14.75; S 11.30.

^{1}H NMR (300 MHz, DMSO/CCl_4, 1/3, 0.83 d (3H, CH_3, 6.3); 0.87 d (3H, CH_3, 6.3);1.06 d (3H, CH_3, 6.0): 1.34-1.63 m (4H); 1.77 ddd (1H, 13.2, 9,6, 4,8); 2.99 m (1H); 3.68 m (1H, OCH); 4.33 d (1H, OH, 5.4); 4.58 ddt (1H, NCH_2, 15.9, 6.3, 1.5); 4.74 ddt (1H, NCH_2, 15.9, 4.8, 1.8); 5.18-5.25 m (2H, =CH_2); 5.89 dddd (1H, =CH, 17.0, 10.5, 6.3, 4.8); 13.30 (1H, SH).

5-(2-hydroxy-7-methyloctan-4-yl)-4-phenyl-2,4-dihydro-3H-1,2,4-triazole-3-thione (7s). Yield 74%, 131-132°C. R_f 0,60 (ethanol: benzene: hexane – 1:3:3). Found, %: C 63.95; H 7.85; N 13.18; S 10.08. $C_{17}H_{25}N_3OS$. Calculated, %: C 63.91; H 7.89; N 13.15; S 10.04.

^{1}H NMR (300 MHz, DMSO/CCl_4, 1/3, 0.79 d & 0.81 d (6H, 2CH_3, 6.6): 0.99 d & 1.04 d(3H, CH_3, 6.0), 0,98 m (2H) ; 1.29-1.59 m (3H) ; 1.78 m (1H); 2.57 m (1H); 3.55 m (1H; 4.04 d (1H, OH, 5.1); 7,31 m (2H, o-C_6H_5); 7.47-7.59 m (3H, m- & p-C_6H_5); 13.48 s (1H, SH).

4-allyl-5-(2-hydroxy-7-methyloctan-4-yl)-2,4-dihydro-3H-1,2,4-triazole-3-thione (7t). Yield 77%, 116-117°C. R_f 0,60 (ethanol: benzene: hexane – 1:3:3). Found, %: C 59.30; H 8.85; N 14.85; S 11.35. $C_{14}H_{25}N_3OS$. Calculated, %: C 59.33; H 8.89; N 14.83; S 11.31.

References

1. *Han MI, Cumaoglu A, Kügükgüzel SG., Marmara Pharm. J.* 2018, 22, 559-569.
2. *Arjun HA, Elancheran R, Manikandan N, Lakshmithendral K, Ramanathan M, Bhattacharjee A, et al.* Front. Chem. 2019, 7, 474.
3. *Afsah EM, Abdel-Galil E, Youssef I, Elsaedy YH.*, American Journal of Chemistry and Materials Science. 2019, 6, 58-65.
4. *Khan T, Yadav R, Gound SS.* An Efficient ,. J. Het. Chem. 2018, 55, 1042-1047.
5. *Dolzhenko AV, Dolzhenko AV, Chui WK.*, Tetrahedron. 2007, 63, 12888-12895.
6. *Gupta M.*, J. Het. Chem. 2007, 44, 1023-1027.
7. *Song YH, Young SH.*, J. Het. Chem. 2010, 47, 1183-1187.
8. *Koparir M, Orek C, Parlak AE, Söylemez A, Koparir P, Karatepe M, et al,* Eur. J. Med. Chem. 2013, 63, 340-346.
9. *Li YH, Zhang B, Yang HK, Li Q, Diao PCh, You WW, et al.* Eur. J. Med. Chem. 2017, 125, 1098-1106.
10. *Zhang G, Hu Y.*, J. Heter. Chem. 2007, 44, 919-922.
11. *Gao F, Wang T, Xiao J, Huang G.*, Eur. J. Med. Chem. 2019, 173, 274-281.
12. *Karnik AV, Malviya NJ, Kulkarni AM, Jaimini D, Jadhav BL.*, J. Heter. Chem. 2006, 43, 489-493.
13. *Arafa WAA.*, J. Heter. Chem. 2010, 47, 1109-1115.
14. *Puthiyapurayil P, Poojary B, Kumar S, Hunnur R.*, J. Heter. Chem. 2011, 48, 998-1005.
15. *Zeydi MM, Montazeri N, Fouladi M.*, J. Heter. Chem. 2017, 54, 3549-3553.
16. *Jian F, Xiao H, Zhu Ch, Xu L.*, J. Heter. Chem. 2006, 43, 925-929.
17. *Song HX, Shi DQ.*, J. Heter. Chem. 2014, 51, 1345-1348.
18. *Federico S, Margiotta E, Salmaso V, Pastorin G, Kachler S, Klotz KN, et al.*, Eur. J. Med. Chem. 2018, 157, 837-851.
19. *Luo YP, Lin L, Yang GF.*, J. Heter. Chem. 2007, 44, 937-943.
20. T.V. Ghochikyan, A. S. Galstyan, Republic of Armenia, patent, N 3442 A // bulletin, 2-05 (2021) 6,87.
21. T. V. Ghochikyan, M. A. Samvelyan, A. S. Galstyan, Republic of Armenia, patent, N742Y//bulletin, AM20220018Y,16-06. (2022)
22. T. V. Ghochikyan, M.A Samvelyan. Republic of Armenia, patent, N 3433 A // bulletin C07D 211/00, 16-02 (2021) 6,87.
23. Tariel V. Ghochikyan, Melanya A. Samvelyan. RA patent, N 3432 A//bulletin C07 C 109/00, 16-02 (2021).

24. Y.S. Nam, J.Y. Park, S.-H. Han & I.-S. Chang, *Biotechnology Letters*. 24 (2002) 2093.
25. T. Akagi, X. Wang, T. Uto, M. Baba, M. Akashi, *Biomaterials*. 28 (2007) 3427.
26. Sh.A. Markarian, H.R. Sargsyan, G.S. Grigoryan, H.H. Ghazoyan, V.V. Chaban, *Journal of Molecular Liquids*. 369 (2023) 120960.
27. M. Hu, Sh. Chujo, H. Nishikawa, Y. Yama-guchi & T. Okubo, *Journal of Nanoparticle Research*. 6 (2004) 479.
28. C. Pascal, J.L. Pascal, F. Favier, M.L.E. Moubtassim, C. Payen, *Chemistry of Materials, American Chemical Society* 141, Published on Web 12/15/1998 – 1999.
29. F. Zhang, S.-W. Chan, J.E. Spanier, E. Apak, Q. Jin, R.D. Robinson, I.P. Herman, *Applied Physics Letters*. 80 (2002) 127.
30. B.J. Berne, R. Pecora, *Dynamic Light Scattering: With Applications to Chemistry, Biology, and Physics,* John Wiley & Sons, Inc. (2000).
31. C.Bellmann, A. Caspari, Fr. Babick, M. Stintz, Ch. Moitzi, *Dynamic and Electrophoretic Light Scattering,* Guidelines for particle-size analysis and zeta-potential determination, (2020) 231.
32. L.K. Shpigun, M.A. Arharova, K.Z. Brainina, A.V. Ivanova, *Analytica Chimica Acta.* 573-574 (2006) 419-426.
33. R.Salazar, M.Pozos, P.Cordero, J.Perez, M.C.Salinas and N.Waksman, *Pharmaceutical Biology*. 46 N3 (2008) 166-170.
34. R.L. Prior, G. Cao, *Free Radical Biology and Medicine.* 27 N 11-12 (1999) 1173-1181.
35. N.F.Santos-Sánchez, R.Salas-Coronado, C.Villanueva-Cañongo and B. Hernán-dez-Carlos, (2019).
36. B. Li, D. A. Pratt, *Free Radical Biology and Medicine*. 82 (2015) 187-202.
37. K.U.Ingold, D.A.Pratt, *Chemical Reviews*. 114 N18 (2014) 9022-9046.
38. X.Colin, *Brief Overview*. Frontiers in Chemistry. 9 (2022) 1-23.
39. K.U. ngold, *Chemical Reviews*. 61 (1961) 563.
40. E.M. Nolan, J. Jaworski, K. Okamoto, Y. Hayashi, M. Sheng, S.J. Lippard, *J. Am. Chem. Soc.* 127 (2005) 16812- 16823.
41. M.J. Rose, N.L. Fry, R.Marlow, L. Hinck, and P.K. Mascharak, *J. Am. Chem. Soc.* 130 (2008) 8834-8846.
42. S.Cretton, L.Breant, L.Pourrez, Ch.Ambuehl, L.Marcourt, S.N.Ebrahimi, M.Hamburger, R.Perozzo, S.Karimou, M. Kaiser, M.Cuendet, and Ph. Christen, *J. Nat. Prod.* 77. N10 (2014) 2304-2311.
43. L.M. Miller, S.C. Mayer, D.M. Berger, D.H. Boschelli, F. Boschelli, L. Di, X. Du, M. Dutia, M. B. Floyd, M. Johnson, C. H. Kenny, G. Krishnamurthy, F.

Moy, S. Petusky, D. Tkach, N. Torres, B. Wu, W. Xu, *Bioorganic & Medicinal Chemistry Letters.* 19 (2009) 62-66.

44. Van Th. Curtius, H. Sauerberg, *Einwirkung yon Hydrazin auf Bntyrolacton-a-carbonsäureesrter Journal für Praktische Chemie.* (Deutsch) 125 N 1 (1930) 139.

45. V.S. Arutyunyan, T.V. Kochikyan, N.S. Egiazaryan, and A.A. Avetisyan, *Arm. Khim. Zh.*, 1-3 (1994) 91-95. Armenian

46. V.S. Arutyunyan, T.V. Kochikyan, E.V. Arutyunyan and A.A. Avetisyan, *https://arar.sci.am/dlibra/publication/201067B: Chemical and Biological Sciences.* 1 (2002) 73. Armenian..

47. A.S. Galstyan, A.I. Martiryan, K.R. Grigoryan, A.G. Ghazaryan, M.A. Samvelyan, T.V. Ghochikyan, V.G. Nenajdenko, *Molecules.* 23 N11 (2018) 2991.

48. A.I.Martiryan, G.A.Shahinyan, V.V.Vardapetyan, *Colloids Interface Sci Commun.* 50 (2022) 100653.

49. B. H.J. Bielski, D.E. Cabelli, R.L. Arudi, A.B. Ross, *Reactivity of HO_2/O_2^- Radicals in Aqueous Solution.* J. Phys. Chem. Ref. Data. 14 (1985) 1041–1100.

50. M.E.Simonsen, J.Muff, L.R.Bennedsen, K.P.Kowalski, E.G.Søgaard, *J. Photochem. Photobiol. A Chem.* 216 N 2-3 (2010) 244-249.

51. *Kochikyan TV, Samvelyan MA, Arutyunyan VS, Avetisyan AA,* Russ. J. Org. Chem. 2003, 39, 1329-1333.

52. Haroutunyan VS, Kochikyan TV, Haroutunyan EV, Avetisyan AA. (2000) [Published in Russian], *Khim. Zh. Arm.(Chemical Journal of Armenia).53*, 121-122.

53. Ghochikyan TV, Galstyan AS, Samvelyan MA., Proc. YSU B: Chem. Biol. Sci. 2023, 57, 68-76.

54. *Bielski BHJ, Cabelli DE, Arudi RL, Ross AB.* Reactivity of HO2/O−2 Radicals in Aqueous Solution. J. Phys. Chem. Ref. Data. 1985, 14, 1041-1100.

55. *Ernestova LS.* Water Quality Indicators. In book "Ecological chemistry of the aquatic environment", M. 1988, 256-277.

56. *Simonsen ME, Muff J, Bennedsen LR, Kowalski KP, Søgaard EG.* Journal of Photochemistry and Photobiology A: Chemistry. 2010, 216, 244-249

57. *Galstyan AS, Martiryan AI, Grigoryan KR, Ghazaryan AG, Samvelyan MA, Ghochikyan TV, et al.* , Molecules. 2018, 23, 2991.

www.ingramcontent.com/pod-product-compliance
Lightning Source LLC
LaVergne TN
LVHW051131210926
848482LV00001B/52

* 9 7 9 8 8 9 9 6 6 1 4 0 2 *